D1249342

N. H. TECHNICAL COLLEGE AT NASHUA
TS155.6.F67 1991
Foston, Arth Fundamentals of computer–i

3 0426 0000 7259 6

Fundamentals of Computer-Integrated Manufacturing

NASHUA COMMUNITY COLLEGE

Fundamentals of Computer-Integrated Manufacturing

Arthur L. Foston
California State University, Fresno

Carolena L. Smith
San Diego State University

Tony Au
California State University, Fresno

PRENTICE HALL
Englewood Cliffs, New Jersey 07632

Library of Congress Cataloging-in-Publication Data

Foston, Arthur L.
  Fundamentals of of computer-integrated manufacturing / Arthur L.
Foston, Carolena L. Smith, Tony Au.
      p.   cm.
  Includes bibliographical references and index.
  ISBN 0-13-333071-0
  1. Computer integrated manufacturing systems.   I. Smith, Carolena
L.   II. Au, Tony   III. Title.
TS155.6.F67   1991
670′.285—dc20   90-37779
                                                              CIP

Editorial/production supervision: Cyndy Lyle Rymer
Cover design: Ben Santora
Prepress buyer: Mary McCartney
Manufacturing buyer: Ed O'Dougherty

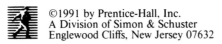©1991 by Prentice-Hall, Inc.
A Division of Simon & Schuster
Englewood Cliffs, New Jersey 07632

All rights reserved. No parts of this book may be
reproduced, in any form or by any means,
without permission in writing from the publisher.

Printed in the United States of America
10 9 8 7 6 5 4 3 2 1

ISBN 0-13-333071-0

Prentice-Hall International (UK) Limited, *London*
Prentice-Hall of Australia Pty. Limited, *Sydney*
Prentice-Hall Canada Inc., *Toronto*
Prentice-Hall Hispanoamericana, S.A., *Mexico*
Prentice-Hall of India Private Limited, *New Delhi*
Prentice-Hall of Japan, Inc., *Tokyo*
Simon & Schuster Asia Pte. Ltd., *Singapore*
Editora Prentice-Hall do Brasil, Ltda., *Rio de Janeiro*

To all individuals who have made, and to those who will continue to make, contributions to the CIM concept, and to all persons, both known and unknown, who helped to develop this book.

Contents

PART 2

Manufacturing Systems Integration
Techniques and Strategies   *225*

Preface

Fundamentals of Computer-Integrated Manufacturing addresses manufacturing as a global closed-loop system comprising four major functions: business, engineering, human resources, and production. The emphasis of this book is on the computer integration of the islands of automation created by isolated computerized systems within these major functions in a manufacturing enterprise. In simple terms, it introduces automation concepts and presents a global concept of the process of product inception, the design process, and all other manufacturing processes, including shipment and customer service. This global approach to manufacturing is based on many technologies that have evolved into a vast information processing and control system that provides many benefits for the manufacturing enterprise, many of which are discussed throughout the book.

This book also serves as an introduction to the subject of computer integration manufacturing (CIM) and provides a comprehensive survey of technical topics related to CIM. It is virtually an encyclopedia on the factory of the future. As a result, the authors assume that the reader has an understanding of the manufacturing process through education or experience.

A unique feature of this book is that it brings many related specialized topics together in a single computer-integrated environment for maximum system efficiency. A nontechnical language is used, and the book contains learning aids such as diagrams, charts, photographic illustrations, acronyms, and definitions of key terms common to the subject. Exercises related to key points are at the end of each chapter. Some chapters include applications of real-world systems to basic concepts.

Fundamentals of Computer-Integrated Manufacturing is designed to be used as an advanced core text in undergraduate programs such as manufacturing, industrial management, industrial technology, and similar programs that provide the students with manufacturing experiences. It is also ideally suited as an entry-level text for graduate-level programs in manufacturing systems integration and in economic and technical issues related to manufacturing computer automation in a school of business and in continuing education programs such as CAD/CAM, CIM, systems analysis and design, and automation technology. The simplicity of the text and its practical approach to systems integration make it particularly useful to those involved in manufacturing management, computer applications and systems engineering, production personnel, and computer technology who desire to become familiar with state-of-the-art flexible automation through systems integration. The book should also be helpful in exposing manufacturing and related personnel to the latest maufacturing technologies.

The treatment in this text of CIM and its building blocks is probably more extensive than most instructors would want to cover in a single semester. As a result, an attempt has been made to present the material in a way that will enable the student to acquire a better understanding of CIM through independent study. Furthermore, the book is designed to fit a variety of course requirements. Some instructors will find that, for any number of reasons, an insufficient amount of classroom time is available to permit an exposition of all topics. As a result, each instructor can select those sections of the book that coincide with the topics included in a particular course. Additionally, this book may serve as a useful reference for other topics not covered in other classes.

Last but not least, this book is aimed at those in industry who are responsible for manufacturing production—on time and on budget—of goods that can be competitive in the marketplace. This book will be helpful to students of industry, computer science, engineering, industrial technology, and business who must be informed about the following:

- Integration of CAD/CAM technologies

- CIM concepts and systems

- CAE (computer-aided engineering) systems

- CAB (computer-aided business) systems

This book is divided in two parts. Part I, *Fundamentals of Product Processes and Operations,* addresses the major building blocks of CIM in six easy-to-read, self-contained chapters. Each chapter stands on its own. As a result, each of these chapters can be covered in a thorough and yet concise manner in the discussion of applications of various subsystems in a traditional manufacturing environment. In these chapters, although certain content materials are redundant, it is necessary so that the reader can comprehend the role of each function in manufacturing.

Part II, *Manufacturing Systems Integration Techniques and Strategies,* delves further into the concept of CIM. It discusses the role of computers in the integration of all phases of product inception; research and design; production planning, controls, and operations; installation; and maintenance of the total manufacturing enterprise, including product support.

Author's Note: Numbers set within brackets in the text (i.e. [1, 1-5] refer to (1) reference number at the end of the chapter and (2) page numbers.

Acknowledgments

We are indebted to many people and organizations for their help and contributions during the preparation of the manuscript. These include:

- Algor Interactive Systems
- Allen-Bradley Company
- Ariel Communications, Inc.
- The Boeing Company
- CalComp, Inc.
- CAM-I, Inc.
- Control Data Corporation
- Electronics America, Inc.
- Elsevier Science Publishers
- EverCAD
- FMC
- General Electric Company
- Hewlett-Packard
- Houston Instrument Division, AMETEK, Inc.
- Hughes Aircraft Company
- IBM
- Intel

- Intermec Corporation
- JDL
- Lowry Computer Systems
- McDonnell-Douglas
- *Manufacturing Engineering Magazine*
- Mazak Corporation
- Mitsubishi Electronics America, Inc.
- Modular Computer System
- Numonics
- OIR Company
- Prime Computer Corporation
- Rust International Corporation
- Society of Manufacturing Engineers
- Sun Microsystems
- Thomas Publishing Company
- Viscom
- Westinghouse Electric Company

Fundamentals
of Computer-Integrated
Manufacturing

PART I

Fundamentals of Product Processes and Operations

Chapter 1 describes the manufacturing process as a closed-loop system within itself and points out that this manufacturing system is made up of *four major* functions. The functional relationship between functions in the manufacturing process is discussed. Basic manufacturing technology developments are defined and traced through various development phases to and including CIM. In addition, manufacturing information flow is discussed in both the manufacturing and production cycles. Chapter 1 forms the hub of the book, and the major functional subject areas are expanded and discussed in greater detail in other chapters.

Chapter 2 establishes the framework for the applications of computers in the manufacturing enterprise's processes. Computerized automation concepts and controls, local area networks and communications, and database systems are discussed as essentials in building and holding together the factory of the future.

Chapter 3 describes the role of the engineering function's systems in the manufacture process. Basic engineering systems are discussed with emphasis placed on computer-aided design (CAD).

Chapter 4 discusses the role of the production function's systems, traditionally called *manufacturing.* The relationship between computer-aided production (CAP) and computer-aided manufacturing (CAM) is discussed along with popular automated production systems. State-of-the-art production technologies are illustrated.

Chapter 5 explores the integration of CAD systems and CAM systems (CAD/CAM) over the complete production cycle of the enterprise. This chapter also includes topics on group technology (GT), flexible manufacturing systems (FMS) and integrated database systems, and integrated communication systems.

Chapter 6 covers the business systems that are a part of the manufacturing environment and how the functional systems are integrated into CIM. Discussions include how computerized business systems, computer-aided business (CAB), help to bring together the total company's resources to support the production function.

1

The Manufacturing Process

Introduction to Manufacturing

Manufacturing is a production business that is intended to make a profit. Profits result from the product development cycle, quality, reliability, uniformity, pricing, public image, productivity, team work, and so on. The primary objective of a manufacturing business is to convert raw materials into quality goods that have value in the marketplace and that are sold at competitive prices. The goods are produced through good management techniques in the use of such resources as capital, human labor, materials, equipment, and energy involving many activities and operations. *Manufacturing* may be defined as a series of interrelated activities and operations involving the design, selection of materials, planning, production, quality assurance, management, and marketing of discrete consumable and durable goods. The interaction among these activities and operations form a total manufacturing system, as shown in Figure 1.1. A *system* is an organized collection of human resources; machines, tools, and equipment resources; financial resources; and methods required to accomplish a set of specific functions. Many processes are used in meeting the primary objective of a manufacturing system in transforming or converting a set of input parameters of new materials and shapes into an output of proper size, configuration, and performance according the specifications for it. This transformation also involves more than physical processes: Management, programs, support, training, consulting, and customization are also involved. Thus, manufacturing is itself an integrated system in action that makes and markets a quality product and makes a profit. Factors contributing to increased profits are improved by the following:

1. System efficiency and throughput
2. Product quality and reliability
3. Productivity at lower product costs
4. Efficient and effective management techniques

Probably the most important of these factors is improved system efficiency and throughput. *Throughput* is the total volume of work performed by the system over a given period, and it has a direct relationship to the other factors. An efficient manufacturing system must be viewed as a system that melds the relationship, interaction, and integration of the individual interrelated activities and operations into a total manufacturing system. That is, a manufacturing system is an enterprise that makes good use of resources to produce products at a profit [1, 6].

Many companies have groups of related activities that seem to act as individual activities. For instance, planning may take place at a secondary level, operating in a standalone mode. As a result, the whole manufacturing system is fragmented. Manufacturing involves not only planning but a series of other interrelated activities such as design, materials selection, production, quality assurance, management, and marketing. For example, design affects materials, planning, production, and so on. Because each group of related activities in the system has a direct bearing on the other groups in the system, a systems planning approach to manufacturing is necessary. A systems planning approach involves analyzing all activities of the system to

Figure 1.1 An Elementary Manufacturing System

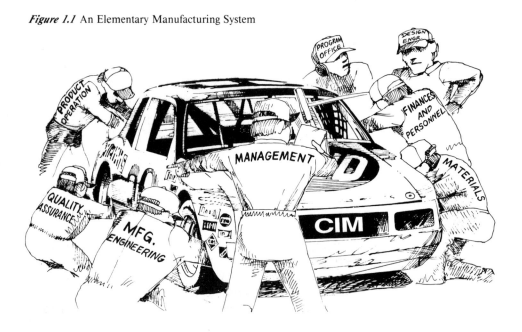

Fundamentals of Product Processes and Operations

determine precisely the functional interrelationships among them and to determine what must be accomplished to make the activities work together efficiently in a systems environment. Such planning also decides how to accomplish the integration process. Some advantages of a system approach follow:

1. Unnecessary duplicated efforts are eliminated.
2. Vital information passes efficiently through the system.
3. The system throughput is improved.
4. Production costs are reduced.
5. Product quality and reliability are improved.
6. Each group knows its relationship to other groups and how it affects the other groups.
7. The groups working as planned units make the whole manufacturing system function more efficiently and productively.

In recent years, manufacturing enterprises have been undergoing many changes to improve the efficiency of their systems in an attempt to produce higher-quality products more economically and to become more competitive in world markets.

Major Manufacturing Functions

A manufacturing enterprise may be viewed as a global system (Figure 1.2A) represented by a small universal set of interrelated functional elements. The functional relationship of each activity to other activities is an important criterion in the design of a manufacturing system. Each activity has an effect on other activities in the total system. As a result, activities are placed in groups to form major functions. A *major function* consists of a group of interrelated activity clusters arranged in terms of an end to be achieved or a behavior required to achieve the end. Under this definition, engineering is a major function. Conceptually, the major functions in a total manufacturing system are business, engineering, human resources, and production (Figure 1.2B). These major functions are further divided into tighter interrelated activities to form special application subsystems. Examples of such divisions of the major functions shown in Figure 1.2B are the following:

- Business Function (business systems): Marketing, inventory management, early planning, production planning and controls, finance, operations management, operational research.
- Engineering Function (engineering systems): Design, drafting/graphic, problem definition, synthesis, analysis and optimization, evaluation, documentation, product review, product presentation.

REPRESENTED BY A SMALL UNIVERSAL SET OF INTERRELATED FUNCTIONAL ELEMENTS

Figure 1.2A Global Representation of Manufacturing

Figure 1.2B Major Functions in a Manufacturing System

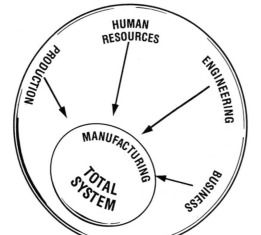

- Human Resource Function (human resources systems): Labor relations, education/training, recruitment, work motivation, management, customer relations, law.
- Production Function (production systems): Fabrication, assembly, inspection and test, materials handling, material monitoring, process monitoring and control, data collection.

A diagram showing an organization structure of the subdivisions in a manufacturing enterprise is illustrated in Figure 1.3. Grouping interrelated functional activities, often called divisions, will vary in name and organization structure from one manufacturing enterprise to another, depending on such considerations as the degree of interaction among subsystems (divisions), types of output goods, methods of manufacturing production, company's philosophy, and functional organization of the enterprise. These relationships are usually shown on an organization chart and described in an organization manual [2, 38]. There is no standard manufacturing organization structure for manufacturing enterprises.

A division may be divided into smaller groups that are often called departments. A functional organization structure of a planning department is illustrated in Figure 1.4. This planning division brings together the planning function of each major function shown in Figure 1.3 under one administrative officer. Other strategies can be applied. Other departments in the manufacturing enterprise can be organized similarly to carry out their operations effectively.

The number of divisions (levels) and names of the levels in a total manufacturing system vary from system to system to form various system configurations and organization structures. Regardless of the organization structure, however, a total systems approach to manufacturing (planning, implementation process, and ongoing manufacturing operations) is essential. An example of a functional organizational chart of a typical manufacturing firm is illustrated in Figure 1.5, which shows the functional structure of an enterprise with certain areas of assigned respon-

Figure 1.3 A Conceptual Organization Functional Diagram

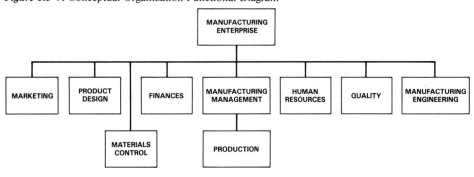

A MANUFACTURING ENTERPRISE'S FUNCTIONAL ELEMENTS MUST BE <u>IDENTIFIED</u>, <u>DEFINED</u>, <u>ORGANIZED</u>, AND <u>CONTROLLED</u> IF REAL BENEFITS OF MANUFACTURING ARE TO BE ACHIEVED.

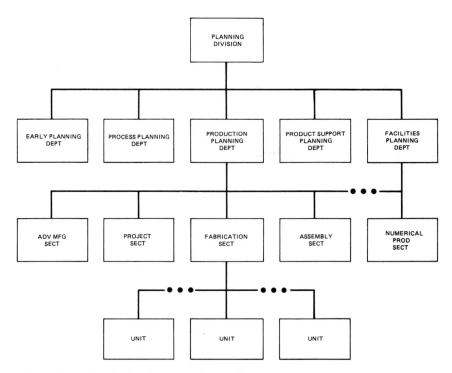

Figure 1.4 An Organization Structure of a Planning Department

Figure 1.5 An Organization Chart of a Traditional Manufacturing Firm

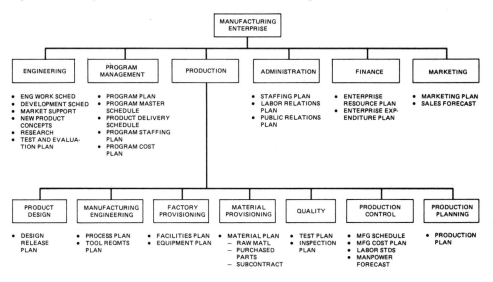

Fundamentals of Product Processes and Operations

sibilities for each function. Each function is divided into subfunctions. An example is production, the basic functions of which are shown in Figure 1.5 with their assigned responsibilities [18]. Because of advances in manufacturing technology, many of the distinctions between functions are disappearing, resulting in other organizational strategies, some of which are discussed in Chapter 7.

Objectives of Organization

The overall objective of a manufacturing organization is to develop teams that function as a single instrument for low-cost production [2, 39]. The manufacturing organization should prove the concept that the whole is greater than the sum of its parts.

Divisional and departmental activities in an enterprise are set up to obtain efficient and maximum performance. It is expected that the firm's managers will control operating results, conserve and coordinate human effort, and optimize costs and profits. From top to bottom, each member of the organization must be motivated to maximize results on the job, to coordinate his or her efforts with those of other departments, and generally to develop the necessary team spirit. Without mutual understanding and cooperation, the organization whole can easily turn out to be less than the sum of its parts.

In summary, a typical manufacturing enterprise can be viewed as representing a conceptual system whose major functions are business, engineering, human resources, and production. Each function in the system is a subsystem entity—that is, a system within a system. The major functions form a total manufacturing system in the sense that they are interrelated in the enterprise's missions and must be coordinated. Coordination is necessary because what is done in any one function has effects on others. An example is that a poorly engineered product will have an impact on the firm's marketing program, which will eventually result in lowered production. The ripple effects will also be felt in human resources. Similarly, personnel policies will have an influence on the workers' morale, and the consequences will be manifest in the production, business, and engineering functions.

Thus, because a manufacturing enterprise can and must be treated as a total system, one can conceptualize that each function may operate independently of the other functions (standalone) under controlled conditions to meet the designed functional requirements. Also, each function must be conceptualized as working collectively with other functions in the environment of a total manufacturing system.

The Manufacturing Cycle

Customer–Marketing/Sales

A manufacturing cycle may be defined as a closed-loop system including all activities and operations from product inception, design, and all activities and operations up to and including product delivery, support, and service (Figure 1.6).

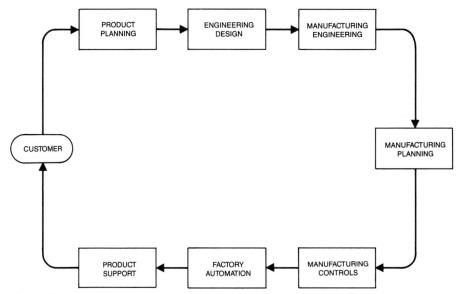

Figure 1.6 A Conceptual Manufacturing Cycle

Figure 1.6 illustrates seven major phases: product planning, engineering design, manufacturing engineering, manufacturing planning, manufacturing controls, factory automation (production), and product support.

The closed-loop system's operations are repeated indefinitely to meet the missions of the enterprise. Such a system uses control techniques during various phases of the production process in which production output processes are fed back for comparison with the product's input requirements. The purpose of the comparisons is to reduce any differences between process input commands and output response.

Manufacturing Phases

The manufacturing cycle shown in Figure 1.6 is exploded in Figure 1.7. The arrangement of the phases in a logical order to produce a product forms a manufacturing cycle.

For simplicity, the manufacturing cycle shown in Figure 1.6 is divided into six major phases by combining phases one and seven, and making the customer the focal point in Figure 1.7.

1. Early planning and product support
2. Engineering and product design
3. Manufacturing engineering
4. Production planning
5. Production controls
6. Production
7. Product shipment and support

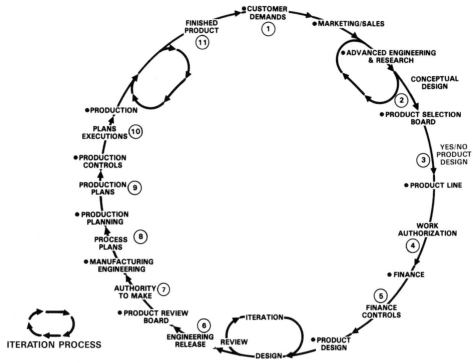

Figure 1.7 A Typical Manufacturing Cycle

Many steps of interrelated activities take place during each of the six phases. They represent the product life cycle over and through all manufacturing processes. The steps in the manufacturing cycle are different, depending on the type of industry, product, size of the company, and management system [3, 16].

Early Planning

Many activities take place during Phase 1. Typical of such activities are forecasting, product selection and contracts, advanced engineering and research, business and product support planning, and operational research. Teams from the major functions work together during this phase. In many cases, developmental models and prototype units are fabricated by production personnel. The resulting product familiarity is readily transferred to production lines. In addition, procurement teams maintain cognizance of all complex articles and major subcontract procurement throughout design, development, and production phases of the product.

Product Demands

The manufacturing cycle generally begins with a need, a mental picture of a product, that in many cases originates with customers' demands or with marketing forecasts, as shown in Figure 1.7. The manufacturing cycle is driven by product demands and marketing forecasts, as shown in Phase 1 of the cycle. These demands, Step #1, are evaluated, and the results are sent to advanced engineering and research for further evaluations. Product research and product requirements are determined. Design options and costs are discussed right away. Studies are conducted to determine product producibility, needed resources, and profits. The output from advanced engineering is a conceptual design, Step #2 in Figure 1.7.

Product Conceptual Models

Product conceptual models are routed to a product selections board (PSB) for review, study, and further evaluations. In many cases, the board requests advanced engineering to participate in the evaluation process. A "yes" decision, Step #3, to manufacture a product is issued through a document such as a work authorization document (WAD), Step #4, by PSB.

The WAD activates finance to provide funding for the various operations needed to produce the product. Funding and accounting are provided through various finance and control documents, Step #5 in Figure 1.7.

Engineering Design

Product Design

Even though product design engineering is actively involved during the early planning phase, it also teams with the production function during the product design phase. As previously discussed, the production function's teams work jointly with design engineering teams. Production provides design engineering with producibility information during product design. Producibility information is concerned with production's fabrication, assembly, test, and inspection capabilities of the designed product. Additional involvement of design engineering is discussed in Chapter 1.

During the joint efforts and iteration processes that have been discussed, design engineers analyze, synthesize, design, and cultivate the product to meet a set of design specifications. If the product as designed meets the specifications, it is approved by engineering management. Product specifications and documentation are released through engineering drawings, lists of materials, and performance specifications. The release is called an engineering released document (ERD), Step #6.

Product Review

The ERD is sent to a product review board, which studies the general suitability of the product in the common marketplace. On the basis of this and other pertinent product information, corporate management makes a product decision. A decision to manufacture the item is issued through such documents as an authority-to-make document, Step #7.

Authority to Make

The authority-to-make document is the initial authorization to the production function. It describes the scope of work explicitly, designates start and stop dates, and specifies a total budget for the task. The document also assigns charge account numbers that provide for cost accumulation and visibility at each level of the work breakdown structure.

Manufacturing Engineering

The authority-to-make document enables manufacturing engineering to start operations. Manufacturing engineering, sometimes called production engineering, develops the production processes by which the design is translated into manufactured goods at the lowest possible cost. Typical functions of manufacturing engineering are: (1) process engineering, (2) tool engineering, (3) industrial engineering, (4) plant engineering, (5) administration and control, and (6) process plans, Step #8. A *process plan* is a detailed plan for the production of a piece, part, or assembly. It includes a sequence of steps to be executed according to the instructions in each step and consistent with the controls indicated in the instructions.

Production Planning and Controls

Authorization-to-make Document

The authority-to-make document of a product is translated into production plans and controls, Step #9. *Production planning* may be defined as the systematic scheduling of workers, materials, and machines by using lead times, time standards, delivery dates, work loads, and other similar data. The purpose of production planning is to produce products efficiently and economically and to meet desired delivery dates. Various activities may also be carried out serially and concurrently during other phases of the manufacturing cycle.

Production control activities, Step #10, involve directing or regulating the movement of goods through the entire manufacturing cycle from the requisitioning of raw materials to the delivery of the finished product, Step #11. Typical production control systems are inventory control, operation schedules, dispatching, expediting, quality control, labor performance, product routing and control.

Production

The production phase of the manufacturing cycle involves the operations of changing the shape, composition, or combination of materials, parts, or subassemblies to increase their value. Many production operations transform raw materials into finished goods. Typical shop floor (the factory) systems are fabrication, assembly, testing, inspecting, and evaluating, material handling, and machine maintenance.

Finished Product

The final phase of the manufacturing cycle is to deliver to the customer quality products and services in a timely manner. This phase also provides complete support for the products. The products should be priced according to market surveys and be of superior quality, reliability, and performance.

The Production Cycle

A production cycle (often called product life cycle) involves interrelated operations that begin with product design and end with product shipment. It is a subcycle within the manufacturing cycle illustrated in Figure 1.7. The production cycle shows the various activities and functions that must be accomplished during the design and production phases illustrated in Figure 1.7. The production cycle is driven by the WAD (Step 4 of Figure 1.7) as shown in Figure 1.8A, and ends with product shipment. Depending on the nature and design of the manufacturing system, however, there will be differences in the way the production cycle is activated.

In some cases, the product preplanning activities are performed by the customer and the product is produced by a different firm. In other cases, design and production are accomplished by the same firm. Whatever the case, the production cycle begins with a design concept, an idea for a product, that is translated into a plan for the product through the design engineering process [4].

For discussion purposes, interrelated operations during the production cycle are also divided into six phases, as illustrated in Figure 1.8B. Some of the phases are performed in sequence, whereas some are performed concurrently with other manufacturing operations. The manner in which the designer's concept moves through various phases of production is shown in the illustrated production cycle (Figure

Figure 1.8A A Conceptual Production Cycle

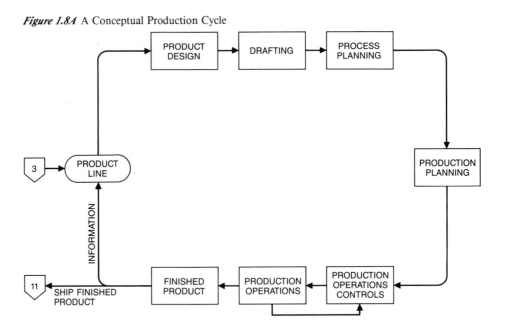

Fundamentals of Product Processes and Operations

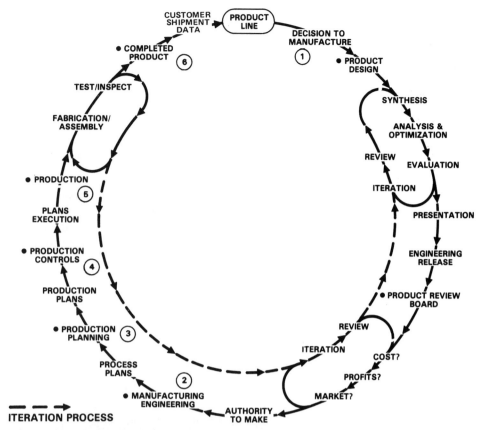

Figure 1.8B A Typical Production Cycle

1.8B). As a result, a production cycle may be viewed as the time it actually takes to produce a product.

As previously stated, phase sequences in a manufacturing firm will vary depending on the philosophy and organization of the company and the size and type of product. This cycle is a subcycle in the manufacturing cycle, Figure 1.7, starting at point 3. The production cycle is completed with information flowing to point 11, Figure 1.8A. The illustrated production cycle is divided into six phases as shown in Figure 1.8B. Starting with a product design Phase 1 in Figure 1.8B, a product cycle is driven by the work authorization document (WAD)—that is, the decision to manufacture. Although the production cycle shown in Figure 1.8B is typical, it is by no means universally applicable. As a result, each manufacturing firm must develop its own version of Phase 1.

Product Design

During product design, Phase 1, the product concept is cultivated, refined, analyzed, and improved. Product engineering has the basic responsibilities of establishing final product requirements, selecting design approaches, choosing the design from alternatives, designing the directive and layout, and building the prototype. The product design is documented by means of component drawings, specifications, and a bill of materials. Product design information is sent to manufacturing engineering, Phase 2.

Manufacturing Engineering

The engineering release document (ERD) activates manufacturing engineering. Some primary responsibilities of manufacturing engineering are production releases, consultation with product design on producibility, process plans, tool engineering, standards, plant engineering, administration and controls, and research. Manufacturing engineering is also highly visible in troubleshooting production problems. Manufacturing engineering does not design products, nor does it take an active role in day-to-day production. Its role is to translate the engineering design into the manufacture of goods and products.

Production Planning

A plan for production is set up through routing, loading, and scheduling, which are the basic planning functions. A master schedule and production control procedures to accomplish a desired result are developed during this phase through a systematic approach to the scheduling of such manufacturing resources as manpower, machines, and materials. Schedules are set by such factors as lead times, time standards, delivery dates, work loads, and similar data for the purpose of producing products efficiently and economically. Schedules are necessary for meeting desired delivery dates and for generating purchase orders, in-shop production schedules, and manufacturing resource plans.

Routing determines where each operation on a part or subassembly will be done. A route sheet gives the order of operations that need to be done to complete a part. After routing is completed, the job is loaded against the particular machine or work area. Loading is the total time it takes to do the needed operations on the part.

Production Controls

Production controls, sometimes called operations management or production management, are the factory's nervous system. This system detects production conditions and problems and sends the necessary directives to keep operations going according to schedule and plans. The schedule and plans are put into action through dispatching through directives. The factory must have directives before it can perform the many operations required to fabricate parts and produce finished products. Production controls send the necessary continuous streams of directives to all parts of the factory telling what to do, when, where, and how many. The processes and operations are monitored to see if any corrective action or preplanning is needed. These are typical control functions. Managing production activities is the responsibility of production control. Many production and control activities are carried out concurrently with production activities in other phases of the production cycle.

Production

Factory floor activities form a vital link with production planning and control in the production processes. The factory floor activities are carefully monitored by the production planning and control functions. These interrelated operations have made production the focal point in the product processes. Production was placed in the center of the manufacturing enterprise model on purpose. The ideas developed by R&D, the human resources obtained by industrial relations, and many other resources purchased with money from financial affairs are all used in the production area to build products. Marketing sees prototypes of the product.

The actual processing of materials, the fabrication of parts, and assembling parts into finished products are the responsibilities of the production function. Many activities are going on during this phase, such as engineering changes and changes in production control directives, parts fabrication and assembly, operations testing and inspections, operations data collection, tracking, and material handling

Figure 1.9 A Typical Product Work Flow in a Production Cycle

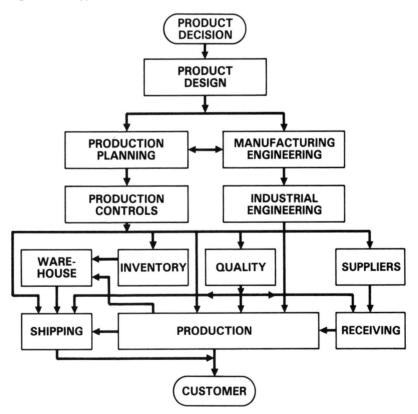

operations. Thus, the production function embraces the implementation of the design, the process, the machines, and the material handling systems to produce a product that is made to all specified requirements.

Finished Product

The product from the production phase moves to the final phase of the manufacturing cycle, as well as in the production cycle, linking the customer and marketing. Typical activities in this phase are packing and boxing, product handling, shipping, and storage. A close relationship exists among the final product, inventory control, marketing, and support services to establish and maintain good customer relations.

A flow diagram of macroactivity sequences for a production cycle is shown in Figure 1.9. A flow diagram is a graphic representation on a flow plan of the work area involved in the location of work stations and the paths of movement of people and materials. The direction taken by personnel or materials as they progress through the manufacturing or processing sequence of events is referred to as a flow line. Included in the activities are quality and inventory controls. In brief, the typical production cycle can be looked on as representing a system within itself, the major functions of which are design, process planning, and production (Figure 1.10).

Figure 1.10 The Big Three in a Production Cycle

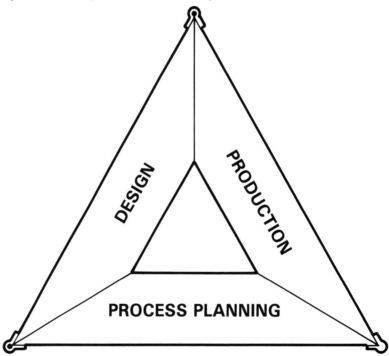

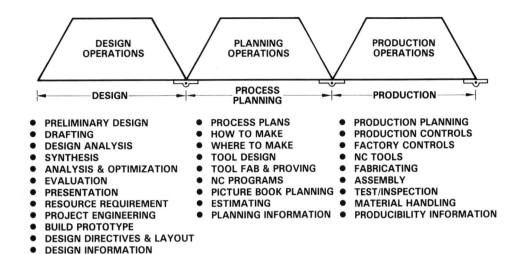

PRELIMINARY DESIGN
- PRELIMINARY DESIGN
- DRAFTING
- DESIGN ANALYSIS
- SYNTHESIS
- ANALYSIS & OPTIMIZATION
- EVALUATION
- PRESENTATION
- RESOURCE REQUIREMENT
- PROJECT ENGINEERING
- BUILD PROTOTYPE
- DESIGN DIRECTIVES & LAYOUT
- DESIGN INFORMATION

- PROCESS PLANS
- HOW TO MAKE
- WHERE TO MAKE
- TOOL DESIGN
- TOOL FAB & PROVING
- NC PROGRAMS
- PICTURE BOOK PLANNING
- ESTIMATING
- PLANNING INFORMATION

- PRODUCTION PLANNING
- PRODUCTION CONTROLS
- FACTORY CONTROLS
- NC TOOLS
- FABRICATING
- ASSEMBLY
- TEST/INSPECTION
- MATERIAL HANDLING
- PRODUCIBILITY INFORMATION

Figure 1.11 Illustrated Activities in a Production Cycle

These three functions jointly form a Big Three production triangle that illustrates the interrelationship of the production cycle functions and the support each offers.

By opening the triangle as illustrated in Figure 1.11, a typical list of production sequences is shown. Through the efforts of qualified people, successful implementation of the triangle will result in the attainment of the basic functions of design, process plans, and production. Hence, it is necessary that the human resource function be concerned with recruitment, selection, training, motivation, customer relations, and management.

Developments in Manufacturing Technology

Overview

Manufacturing technology has been around for many years. Over these years, it has gone through many changes, ranging from the simple to the complex. The driving forces behind the changes were people's desires to improve basic needs such as food, clothing, shelter, and recreation. To meet these desires, methods have been developed for producing simple devices such as weapons for obtaining food to today's modern manufacturing systems, which use computers to produce such items as televisions and space vehicles [6, 1].

People continue to seek ways to produce devices that will improve their standard of living, meet their basic needs, and control their environment. Such improved methods create demands that translate to larger quantities of devices that are produced more quickly than previously and that are of better quality at a lower cost per item. Thus, developments from elementary manufacturing processes for simple devices to advanced automated systems and complex manufacturing processes have evolved through several stages of development that are the results of ideas, combinations of various resources, and hard work by many people.

The construction and application of simple machines for production started in Europe around 1770 [7, 1]. These developments moved production from the home to factories, marking the beginning of the Industrial Revolution. Mechanization signified the movement from making products by hand in the home to making products by machines in factories.

Mechanization also created a system of mass production, which placed a demand on machines to duplicate parts with a high degree of accuracy. This resulted in a need for more accurate measuring tools, improved measuring techniques, and standards to help manufacturers make interchangeable parts. Fixed automation mechanisms and transfer lines were major results of mass production. A *transfer line* is an organization of manufacturing facilities for faster output and shorter production time. Mass production was the first example of automated production.

Fixed automation gave way to machine tools with simple automated controls. This type of controller operates automatically to regulate a controlled variable or system. Advancement in controller technology opened the new era of automation, called *programmable automation.*

Programmable automation is designed to accommodate changes in a product. A new technique in automation, numerical control (NC), developed around 1952 [7, 2]. Numerical control is based on digital computer principles, a form of programmable automation that controls manufacturing processes by numbers, letters, and symbols [3]. Some applications of NC are found in such areas as training, milling, inspections, welding, drafting, materials handling, and assembly. Advances in computer technology extended NC to direct numerical control (DNC), computer numerical control (CNC), graphical numerical control (GNC), and voice numerical control (VNC). These topics are discussed in Chapter 4.

Numerical control caused a revolution in the manufacture of discrete metal parts. The success of NC led to a number of extensions such as adaptive control (AC) and industrial robots. Adaptive control determines the proper speeds and feeds during machining as a function of variations in such factors as hardness of work material, width or depth of cut, air gaps in the part geometry, and so on. It denotes a control system that measures certain output process variables and uses these to control the speed and feed of the machine tool. Typical process variables used in AC machining systems are spindle deflection or force, torque, cutting, temperature, vibration amplitude, and horsepower. Industrial robots started playing a major role in manufacturing during the late 1970s. Initially, robots were used for material handling. Today's robotic technology, however, has been developed to such an extent that robots are used to perform many high-level tasks in manufacturing.

Computers are being given an increasingly important role in manufacturing systems. A computer's ability to receive and handle large amounts of data, coupled with their fast processing time, makes a system approach indispensable. The use of computers in manufacturing is now coming of age. Computer application in manufacturing production controls the physical process and is typically referred to as computer-aided manufacturing (CAM). It is built on the foundation of such systems as NC, AC, robotics, automated guided vehicle system (AGVS), automated storage/

retrieval system (AS/RS), and flexible manufacturing system (FMS). Some of the new uses are briefly discussed below. More detailed discussion is presented in subsequent chapters.

Computer-Aided Manufacturing

Computer-aided manufacturing is the effective use of computer technology in the planning, management, control, and operations of a manufacturing production facility through either direct or indirect computer interface with physical and human resources of the company [3], and it is described in Chapter 4. Computers play important roles in CAM systems [8]. They assist with operations those in the production leg of the Big Three triangle (Figure 1.11), and they also integrate manufacturing data for these operations into a common database. A database is an all-user data set logically associated with a single conceptual scheme as well as with its associated undefined data sets. Database management concepts are applied to CAM operations to speed up data access and to ensure that all users work from a common design. The use of computers in manufacturing operations has been growing rapidly in the production process, also. Under the CAM definition, computer applications include such systems as inventory control, scheduling, machine monitoring, and management information. These applications are primarily for transferring, interpreting, and keeping track of manufacturing data [9].

Data communication is the glue that holds production operations together. It is the factory's central nervous system that supports the six vital information, coordination, and control functions: distributed control, monitoring, data acquisition, supervisory control, program support, and management information.

Computers have been used in the product cycle to facilitate the design process since the early 1960s. When computers are so used, the process is referred to as a computer-aided design (CAD). It is a broad subject addressing applications in many activities of the design function.

Computer-Aided Design

In summary, CAD is a process that uses computers to assist in the creation, modification, analysis, or optimization of a design (see Chapter 3). It refers to the integration of computers into design activities by providing a close coupling between the designer and the computer (Figure 1.12). Typical design activities involving a CAD system are preliminary design, drafting, modeling, and simulation. Such activities may be viewed as CAD application modules interfaced into a controlled network operation under the supervision of a computer. Additional CAD involvements are illustrated in the design leg of Figure 1.11.

Computer-Aided Process Planning (CAPP)

One of the most important functions of manufacturing engineering (Phase 2 of the production cycle in Figure 1.8) is process planning. Process planning is responsible for detailed plans for the production of a part or an assembly. This

Figure 1.12 A Designer Using a CAD Workstation

operation includes a sequence of steps to be executed according to the instructions in each step and consistent with the controls indicated in the instructions. Closely related to the process planning function are the functions that determine the cutting conditions and set the time standards. These traditional manual and clerical operations have been enhanced by developments in technology. Process planning now is aided by computer-aided process planning (CAPP). The foundation of CAPP is group technology (GT), which is the means of coding parts on the basis of similarities in their design and manufacturing attributes. A well-developed CAPP system can reduce clerical work in manufacturing engineering and provide assistance in production.

Computer-aided process planning systems consist of computer programs that allow planning personnel interactively to create, store, review, edit, and print fabrication and assembly planning instructions. Such systems offer the potential for reducing the routine clerical work of manufacturing engineers. The areas in which they are used are shown in the process planning leg of Figure 1.11.

There are two approaches to CAPP: the variant and the generative. The *variant* systems use parts classification and coding and group technology as a foundation. The parts produced in the plant are grouped into families, each of which is distinguished by its manufacturing characteristics. *Generative* CAPP systems use the computer to create an individual process plan from point zero by employing a set of algorithms to progress through the various technical and logical decisions toward a final plan for production. This process is done automatically without human assistance.

As computers have continued to have an impact on manufacturing operations, engineers and computer scientists have been developing new applications. One of the newest and most important uses for computers is in computer-aided design/computer-aided manufacturing (CAD/CAM).

CAD/CAM

Automated islands exist over the complete production cycle illustrated in Figure 1.8B. Automated islands were created as a result of each function automating its operations in Figure 1.8B. These automated operations continued to work in an isolated environment without using the computer to pass information between the various functions. A key factor in improving productivity is to integrate and control the data flow to and from the automated islands. The integration process of CAD, CAPP, and CAM is referred to as CAD/CAM [10]. Thus, CAD/CAM is the technology concerned with the use of computers to perform certain functions in design, process planning, and production to improve productivity. It is a philosophy of computer-integrating islands of automation and production activities over the complete production cycle by controlling and sharing data among the islands (see Chapter 5).

There are many types of CAD/CAM systems, all of which store, retrieve, manipulate, and display graphical information—all with unsurpassed speed and accuracy. Because the information is stored in computer memory instead of hard copy, the transfer of data tends to be quicker, more reliable, and less redundant. As a result, more can be accomplished in a given time by engineers. Also, product quality and yield are improved, optimizing the use of energy, materials, and manufacturing personnel.

The computer-aided Big Three representation of the production cycles is shown in Figures 1.13A and 1.13B. A typical example of the role CAD, Computer-Aided Design Drafting (CADD), CAPP, computer-integrated production management (CIPM), and CAM play in the production cycle is illustrated in Figure 1.13A. Quality assurance is reflected in all activities over the complete production cycle. The various production activities are imbedded in the triangle of Figure 1.13B. This triangle is not a replacement for the triangle shown in Figure 1.10. It is an overlay on virtually all the activities and functions of the production cycle as illustrated in Figure 1.11. Typical involvements of the Big Three in CAD/CAM are illustrated in Figure 1.14 and in the impact of CAD/CAM activities in a production cycle. A typical sequence of some of the important activities is shown in Figure 1.15, which illustrates a more detailed list of the activities in the production cycle shown earlier in Figure 1.8.

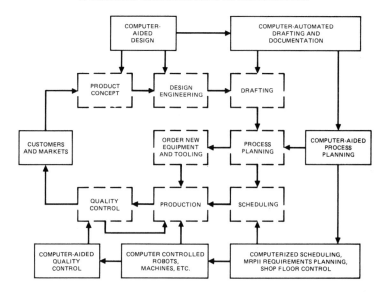

Figure 1.13A CAD/CAM Integration—Computer Assistance to the Production Cycle

Figure 1.13B The Big Three in the CAD/CAM Cycle

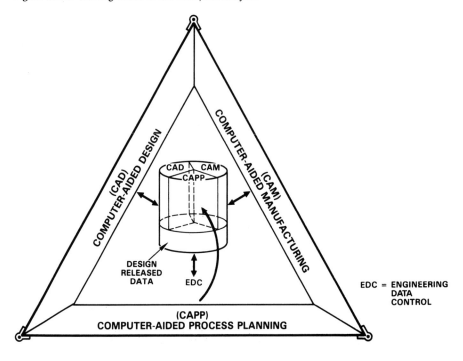

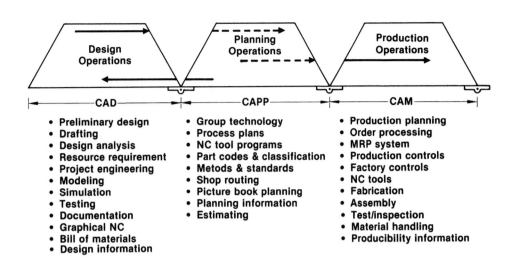

Figure 1.14 Typical CAD/CAM Activities in a Production Cycle

Figure 1.15 A Simplified Sequence of CAD/CAM Activities

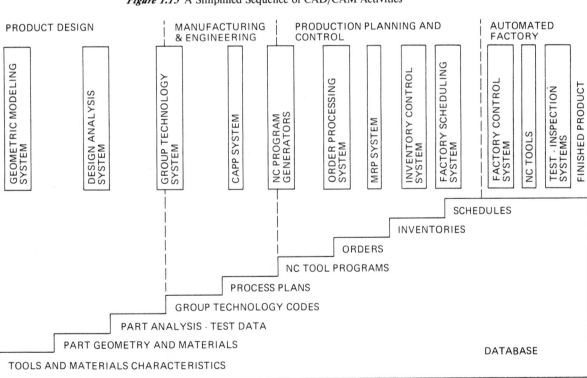

The ever-increasing power of the computer provides for greater integration of manufacturing activities. Systems in the Big Three work to generate a common database such as that shown in Figure 1.13B. Such a database linking operations together is also illustrated in Figure 1.15. Notice that value is added to the data as they move around the cycle. The use of computers to integrate all operations in a manufacturing enterprise will no doubt rise dramatically during the 1990s. This new technology will address manufacturing as a system within itself. In this text, the discussion of the new technology addresses the manufacturing enterprise as a large information system. That is, it addresses computer integration of manufacturing operations and activities over the complete manufacturing cycle (Figure 1.6), from business planning to design, through production, marketing, and product support.

Computer-Integrated Manufacturing (CIM)

Overview

Computer-integrated manufacturing means different things to different people and different industries. However, regardless of the various definitions, CIM represents computers in manufacturing whether they be used early in the product development phase feeding production, factory floor controls, flexible manufacturing systems, and work cells, or used as an information system moving data throughout the manufacturing enterprise. Such an information system concept permits all aspects of the enterprise to have access to timely and accurate information that will help them do their jobs better, improve product quality, and increase productivity.

Computer-integrated manufacturing as a term has undergone a discernible evolution since its coinage by the late Joseph Harrington, Jr., in the 1973 publication of his book *Computer Integrated Manufacturing* [11, 36]. Its definition has changed with time, and industry now perceives it differently. Several definitions are discussed in Chapter 7.

A functional model of a traditional computer-assisted manufacturing enterprise is shown in Figure 1.16. This enterprise is divided into four major computer-assisted functions, each of which, although separated from the other functions by walls, carries out its operations according to its mission.

A manufacturing enterprise was traditionally divided and fragmented into many departments and activities. In many cases, the enterprise was inflexible because communications between departments was practically nonexistent. These conditions resulted in many manufacturing data errors, long production runs, and large inventory, all of which made it difficult to respond to changing market needs. To improve system efficiency, many industries started moving toward integration of the diverse elements of the departments in the manufacturing system.

Each of the four functions shown in Figure 1.16 uses computers and has programs to process data according to its own needs. In many cases, each function uses computers to integrate the many islands of automation of its subfunctions. As a result, computer technologies lead to the development of computer-aided functions

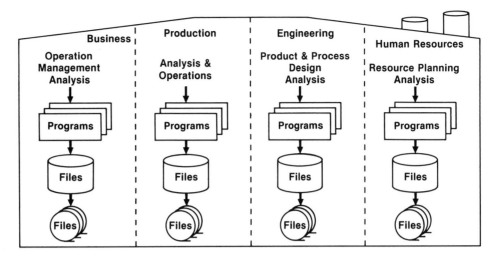

Figure 1.16 Walls Exist Between Functions in a Traditional Manufacturing
Enterprise

such as computer-aided business (CAB), computer-aided engineering (CAE), computer-aided production (CAP), and computer-aided human resources (CAHR). In discussions of systems, CAP is used interchangeably with CAM and carries the same meaning.

Figure 1.16 shows that the need for techniques to pass data between major functions is not apparent. Lack of such techniques, however, will cause many problems—for example, additions, deletions, and changes—during manufacturing (Figure 1.17). Many errors can be generated during these processes that will lower the enterprise's productivity factor. However, an analysis of the manufacturing data used by each major function reveals that certain data are common to all functions (Figure 1.18). The common data, Item 1, can be placed in a common database accessible to all functions. Likewise, Item 2 represents data common to human resources, engineering, and production; Item 3 represents data common to human resources, business, and production; and so on. These common data elements play important roles in the design of database systems. There is, then, no need to keep redundant data, which are prone to errors, in each of the various functions.

The degree of data overlap and interactions among functions depends on the structure of the manufacturing system. Regardless of the structure, however, related applications in the system use the same data. As a result, using a computer to make data available to related applications in the total system is a positive step toward improving the system's throughput and productivity (basic output/basic labor input). Therefore, in a modern manufacturing system, there should be an approach to system design that will make the system effective, efficient, and able to provide data interactions to related applications, programs, and users. Because related applications and activities use some common data, a logical approach to such a system design is to eliminate redundant data wherever possible. One method is to join various files together with a computer through a common database system (Figure

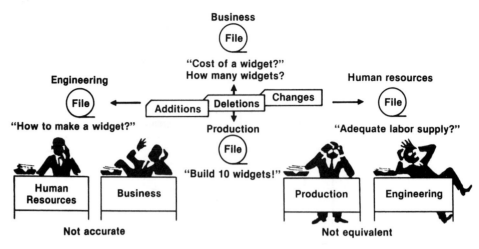

Figure 1.17 Redundant Data Cause Manufacturing Problems

1.19). This concept, called computer-integrated manufacturing, is rapidly becoming a reality in many plants today (see Chapter 7).

The use of computers in integrating many phases of manufacturing will rise dramatically during the next decade. Total systems involving cell control, center control, and area control will be coordinated through factory data systems. The increasing power of the computer will provide for greater local control while sup-

Figure 1.18 Related Activities Use the Same Data

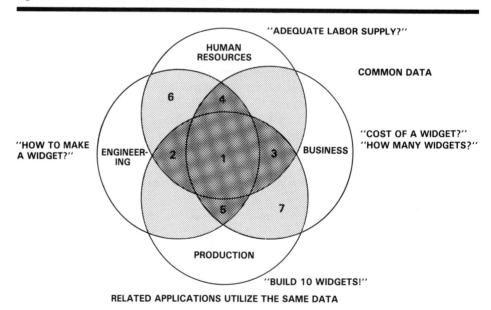

Fundamentals of Product Processes and Operations

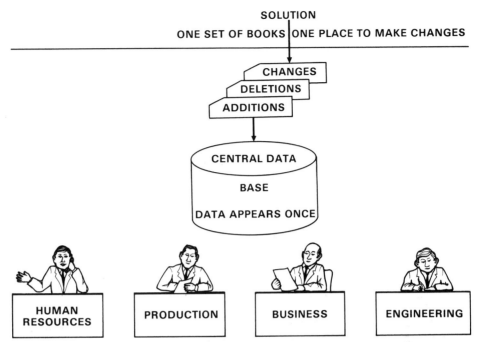

Figure 1.19 A Common Database for a Manufacturing System

plying input to a common data base. The computer will also play a major role in managing the manufacturing cycle and reduce the manufacturing process cycle drastically.

Computer-Integrated Manufacturing (CIM) Defined

Gabor Koves [13] defines CIM as the conceptual basis for integrating the applications and information flow of product design, production planning, and plant operations. The focus of CIM is on information as the crucial element linking all facets of the product enterprise. The computer manufacturer, IBM, sees CIM as being a broad-based computer-integrated manufacturing operation. As IBM defines CIM, it encompasses and provides the conceptual basis for integrating the applications and information flow of the following areas: product design (CAD/CAM), production planning and control (COPICS and MAPICS), and plant operations (industrial automation) [14, 4].

MAPICS (Manufacturing Accounting and Production Information Control System) is ideal for first-time computer users who manage small plants, or for remote facilities of larger organizations. COPICS (Communications Oriented Information and Control System) is for larger companies. It consists of several integrated modules that address specific manufacturing areas that tie production to engineering.

On the basis of IBM's definition of CIM, Koves appears to support the concept in a summary definition: "CIM can also be defined as the integration of *informa-*

tion-generating and information-using systems" [18]. Computer-integrated manufacturing refers to complete systems in which every aspect of the manufacturing process has been computerized and tied into a single controlling, coordinating system. All applications of CAD and CAM, including management planning and control, are in the manufacturing process—from management decisionmaking to shop floor management—which is accomplished through shared access to the system's database [14].

The definition of CIM has changed over time from computerized workcells, flexible manufacturing systems, large-scale automation, CAD/CAM, interfacing and communications concepts, to a "more" state of maturity: CIM as an information system that controls data flow throughout the engineering and manufacturing functions. More recently, with the realization that CIM must be a part of an even larger integration effort, the definition has come to embrace the whole corporate organization so that manufacturing can take its rightful but long-denied place as an equal of marketing and finance in the development and execution of the business strategies that determine a company's future [11, 36].

The Computer and Automated Systems Association (CASA) of the Society of Manufacturing Engineers (SME) defines computer-integrated manufacturing (CIM) as a system that provides computer assistance to all business functions within a manufacturing enterprise—from product design and order entry to product shipment [15]. The term refers to the *total* manufacturing enterprise, *not* to the manufacturing function within that enterprise.

Computer-Aided Manufacturing–International, Inc. (CAM–I) has developed plans for integrated subsystems. CAM–I refers to the enterprise information system concept known as computer-integrated enterprise (CIE), which ties together the seemingly diverse threads of the entire enterprise by integrating the business activities, including product design, manufacturing engineering, the factory floor, and the various support functions [16].

We must consider CIM to be a closed-loop feedback information system whose prime inputs are product requirements and concepts and whose primary output are finished products. True CIM requires a combination of computer software and hardware for product design, production planning and control, and production processes. It encompasses many automated support subsystems, as shown in the total manufacturing cycle in Figure 1.6. Each automated subsystem is referred to as a CIM system (CIMS). The backbone of the totally integrated CIMS enterprise is CIM.

General Electric's concept of a factory with a future is illustrated in Figure 1.20 [17]. This model permits the integration of the many diverse elements in the manufacturing enterprise. Through this model, data are shared by the major functions of the system. A complete CIM implementation results in automating the data flow in the organization from order entry through every step in the process to shipping the finished goods and providing services to the product in the field. Thus, CIM is a vital management system for information, time, people, money, machines, and materials.

General Electric's view of the automated factory integrates elements in Figure 1.20 under the control of, and networked with, computers. Computer-aided engi-

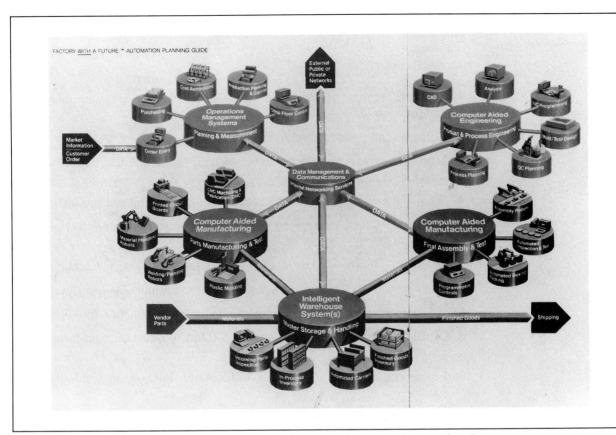

Figure 1.20 General Electric Company's Concept of a Factory of the Future

neering (upper right) contributes information such as machine tool programs and design. Computer-aided manufacturing, integrating such devices as numerically controlled machine tools and robots, automatically manufactures, assembles, and tests. An automated warehouse controls and moves inventory. Operations management functions, such as purchasing and cost accounting, receive input from the computer that is based on activity in the factory.

Both CIM and CIE carry the same meaning in this textbook—that of a manufacturing computer information system. This information system's goal is to reproduce all the product, process, management, and business functions in terms of computer data that can then move freely and be of use throughout the system [11]. The computerization of all operations and activities over the complete manufacturing cycle is referred to in this textbook as computer-integrated manufacturing (CIM). This concept is discussed in Chapter 7.

Under computer manufacturing control, not only are machines linked to machines, but to all related support groups including design, materials handling, assembly, and packaging and shipping. All phases of the manufacturing process become integrated. The location and status of everything in the plant is traceable at the touch of a few computer keys [12].

The Manufacturing Process

31

Exercises

1. Define (a) manufacturing, (b) production, (c) system, (d) system integration, and (e) operations.

2. Name several types of production industries.

3. Name and explain the differences between the four major functions of a manufacturing enterprise.

4. What is the relationship, if any, between manufacturing and production?

5. Explain the difference between a manufacturing cycle and a production (product) cycle.

6. Define (a) CIE, (b) CAB, (c) CAE, (d) CIM, (e) CAD, (f) CAM, and (g) CAPP.

7. What are the Big Three operational functions in a manufacturing cycle?

8. What are the Big Three operational functions in a production (product) cycle?

9. Explain the differences, if any, between CAM, production, and factory floor activities.

10. List several basic subsets of each of the major functions in a manufacturing system.

11. Why is it said that the production function is a system?

12. Is the production function independent of the business, engineering, and human resource functions? Explain.

Self-Study Test

1. Manufacturing is a generic term (encompassing such major functions as business, management, engineering, marketing, and production) relating to the making or producing of goods by using material, machine/tool, and human resources.
   True   False

2. Manufacturing production technology may be viewed as a "productive system" consisting of a set of components whose function is to transform a set of "input" parameters into some desired "output."
   True   False

3. Probably the most important part of a manufacturing system is the relation of the individual parts to the whole.
   True   False

4. Manufacturing production is a transformation process in which raw materials are converted into goods that have value in the marketplace.
   True   False

5. Factory management may be defined as a closed-loop communications network involving scheduling "N" jobs across "M" machines.
   True   False

6. A manufacturing cycle may be defined as an open-loop system including all activities and operations from product inception to delivery and product service.
   True   False

7. Some primary responsibilities of manufacturing engineering are product design, production (authority-to-make) releases, process plans, and tool engineering.
   True   False

8. Production planning involves forecasting the demand for the firm's products and services, and translating this forecast into its equivalent demand for various factors of production.
   True   False

9. CIM consists of modular subsystems, each controlled by computers that are interconnected to form a distributed computer system.
   True   False

10. The impact of CAD/CAM is manifest in all the different activities in the manufacturing cycle.
    True   False

11. The foundation of CIM is the computer.
    True   False

12. It is highly desirable that the manufacturing engineering department provide advice to the product design department on producibility.
    True   False

13. A decision by engineering management that a design is approved and ready for production is made known through an authority-to-make release.
    True   False

14. Group technology involves the grouping of parts, machines, or processes by some criteria to increase technology developments.
    True   False

15. CAD/CAM increases productivity and improves product quality even though it increases production cost.
    True   False

16. The transformation process steps in production, each step bringing the materials closer to the desired final state, are referred to as_____.

1. Production activities
2. Major functions
3. Basic functions
4. Production operations
5. Production processing

17. Two general classifications of production operations required to transform raw material into finished product are (1) _____ and (2) _____.

1. Basic, secondary
2. Fabrication, assembly
3. Basic, finish
4. Design, production
5. Processes, activities manufacturing technology

18. _____ is the factory's nervous system. It sends the necessary continuous stream of directions to all parts of the factory.

1. Process planning
2. Production planning
3. Routing
4. Scheduling
5. Production controls

19. _____ is a transformation process in which raw materials are converted into goods that have value in the marketplace.

1. Manufacturing
2. Production
3. Product design
4. Items 2 and 3
5. All these.

20. Two general types of industrial firms, depending on the nature of their operations, engaged in production are _____ and _____.

1. Manufacturing, process
2. Job shop, mass
3. Job shop, batch
4. Production, process
5. All these.

21. _____ may be defined as the effective use of computer technology in the management, control, and operations of the manufacturing facility through either direct or indirect interface with the physical and human resources of the company.

   1. CAD
   2. CAM
   3. CAP
   4. CAPP
   5. CIM

22. Typical examples or subsystems of the business functions are:

   1. Marketing, finance, product evaluation, production planning
   2. Production planning and controls, finance, marketing, early planning
   3. Product presentation, law, operations management, production planning and controls
   4. Items 2 and 3.
   5. All these.

23. _____ is concerned with the activities that are involved in the preparation of "route sheets" and generate route sheets. Route sheets list the sequence of operations and machines required to produce a workpart.

   1. CAD
   2. CAM
   3. CAP
   4. CAPP
   5. CAD/CAM

24. Typical major functions of a modern manufacturing system are:

   1. Designing, operating, controlling, updating
   2. Manufacturing, transportation, retailing, storage
   3. Engineering, production, business, managerial
   4. Managerial, operations, production, controls
   5. None of these.

25. Typical major phases of a manufacturing cycle are:

   1. Customer demands, conceptual designs, process plans, finished products
   2. Early planning, product design, production planning, production
   3. Work authorization, engineering release document, authorization to manufacture, finished product
   4. Items 1 and 2.
   5. All these.

26. _____ may be defined as the effective use of computer technology to automate operations and integrate "absolute" redundant data into a common data base. It makes the data available to CAD, CAPP, CAM, and support group users during the production phase of the manufacturing process. It also provides manufacturing services to production.

1. CAP
2. CAD/CAM
3. CIM
4. Items 2 and 3.
5. All these.

27. _____ can be described as any design activity that uses computers to assist in the creation, modification, analysis, or optimization of a design.

1. CAD
2. CAPP
3. CAM
4. CAD/CAM
5. CIM

28. The _____ involves interrelated operations, beginning with product creation and ending with product shipment.

1. Manufacturing cycle
2. Production cycle
3. CAD/CAM
4. CIM
5. All these.

29. Which of the following are the three basic functions (the Big Three) of manufacturing production?

1. Marketing, materials, and management
2. Design, instruction, and quality
3. Design, process planning, and production
4. Human resources, management, production
4. Management, marketing, and production

30. Automation is a key to:

1. A shorter workweek
2. Safer working conditions for the worker
3. Lower prices and better product
4. Increasing our standard of living
5. All these.

31. The production function embraces the implementation of the _____ to actually produce a specified product that meets all design requirements.

1. Designs and processes
2. Machines and materials handling systems
3. Planning and processes
4. Materials and processes
5. All these.

32. The implementation of _____ results in the automation of the information flow in a business organization . . . from entry of an order to shipment of the finished product.

1. CAB
2. CAE
3. CAM
4. CAD/CAM
5. CIM

33. _____ will provide the technology base for the computer-integrated factory of the future.

1. CAB
2. CAD
3. CAM
4. CAPP
5. CAD/CAM

34. _____ prepares a listing of the operation sequence required to process a particular product or component.

1. CAD
2. CAP
3. CAPP
4. CAD/CAM
5. CIM

35. _____ planning involves forecasting the demand for the firm's products and services and translating this forecast into its equivalent demand for various factors of production.

1. Early
2. Process
3. Production
4. Work
5. All these.

36. Compare the activities in the production cycle with the manufacturing cycle in CIM:

1. They are used interchangeably
2. The production cycle is a subset of the manufacturing cycle
3. The manufacturing cycle is a subset of the production cycle
4. The manufacturing cycle includes all business functions from marketing, through the production cycle, to product shipment and product support
5. Items 2 and 4.

37. _____ provides computer assistance to all functions in the production cycle.

1. CAD
2. CAM
3. CAD/CAM
4. CIM
5. Items 3 and 4.

38. _____ can be defined as the use of computer systems to plan, manage, and control the operations of a manufacturing production plant through either direct or indirect computer interface with the plant's production resources.

1. CAD
2. CAM
3. CAPP
4. CAD/CAM
5. All these.

39. Computer process control involves:

1. Direct computer interface with the manufacturing process.
2. Monitoring capabilities of the manufacturing process
3. Controlling of the manufacturing process
4. Items 1 and 2.
5. All these.

40. The actual conversion of raw materials and shapes to end products of proper size, configuration, and performance specifications is accomplished by the process of _____ technology.

1. Computer
2. Engineering
3. Industrial
4. Manufacturing
5. Production

41. The *most* important factor(s) in determining the I/O equipment needs of an information system is (are) the

1. Hardware available
2. Rules governing proper mix
3. Information needs of the users
4. Data needs of the organization

42. Raw factory data take on real value to management after they have been:

1. Examined
2. Analyzed
3. Compared
4. Classified
5. All these.

43. Product designs are documented by means of:

1. Component drawings
2. Specifications
3. Bill of materials
4. Items 1 and 2 above
5. All these.

44. The sequence of events that take place during the creation of the product from its beginning until the time the product is shipped is often referred to as _____.

1. A manufacturing cycle
2. A production cycle
3. Production control cycle
4. Flow cycle
5. None of these.

45. Manufacturing _____ is the defining of the requirements and processes that will most efficiently produce a designed part.

1. Controls
2. Engineering
3. Operations
4. Planning
5. Tooling

46. The sequence or individual steps in a transformation process, by which raw materials are converted into goods that have value in the marketplace, are referred to as production _____.

   1. Processes
   2. Activities
   3. Functions
   4. Operations
   5. Jobs

47. The _____ type of manufacturer takes the natural resources and transforms them into raw materials used by other industrial manufacturing firms.

   1. Basic producer
   2. Converter
   3. Fabricator
   4. Manufacturing technology
   5. All these.

48. _____ is the system in which the computer develops process plans for a part that are based on information stored in a data base.

   1. CAP
   2. FIS
   3. MIS
   4. MMS
   5. None of these.

49. An automated flow line consists of several _____ linked together by work-handling devices that transfer parts between stations.

   1. Machines
   2. Work stations
   3. Raw materials
   4. Items 1 and 2.
   5. All these.

50. The objectives for the use of flow line automation are:

   1. Reduce labor costs, integration of operations, and increase production rates
   2. Minimize distance moved between operations, integration of operations and reduce work in progress
   3. Specialization of operations, integration of operations, and minimize distances moved between operations
   4. Reduce labor costs, minimize distances moved between operations, and increase production rates
   5. All these.

ANSWERS TO SELF-STUDY TEST

1.	False	**11.**	True	**21.**	2	**31.**	5	**41.**	3
2.	True	**12.**	True	**22.**	2	**32.**	5	**42.**	5
3.	True	**13.**	False	**23.**	4	**33.**	5	**43.**	5
4.	True	**14.**	False	**24.**	5	**34.**	3	**44.**	2
5.	True	**15.**	False	**25.**	4	**35.**	1	**45.**	2
6.	False	**16.**	4	**26.**	4	**36.**	5	**46.**	4
7.	False	**17.**	2	**27.**	1	**37.**	5	**47.**	1
8.	False	**18.**	5	**28.**	5	**38.**	2	**48.**	5
9.	True	**19.**	2	**29.**	3	**39.**	5	**49.**	5
10.	False	**20.**	4	**30.**	5	**40.**	5	**50.**	5

References

[1] Wright, R. Thomas, *Manufacturing Materials Processing, Management, Careers.* Goodheart-Wilcox, South Holland, IL, 1984.

[2] Amrine, Harold J., Ritchey, John A., and Noodie, Colin L., *Manufacturing Organization and Management,* 5th ed. Prentice-Hall, Englewood, NJ, 1987.

[3] Grover, Mikell P., *Automation, Production Systems, and Computer-Aided Manufacturing.* Prentice-Hall, Englewood, NJ, 1980.

[4] Grover, Mikell P., and Zimmers, Emory W., *CAD/CAM Computer-Aided Design and Manufacturing.* Prentice-Hall, Englewood, NJ, 1984.

[5] Wright, R. Thomas, *Manufacturing Materials Processing, Management, Careers.* Goodheart-Wilcox, South Holland, IL, 1984.

[6] Jambro, Donald J., *Introduction To Manufacturing.* Delmar, Albany, NY, 1982.

[7] Koren, Yoram, *Computer Control of Manufacturing Systems.* McGraw-Hill, New York, 1983.

[8] Hollingum, Jack, "Computers—The All-Bracing Route to Computer-Aided Manufacturing." *The Engineer,* October 26, 1978.

[9] Hordeski, M. F., *CAD/CAM Techniques.* Reston, Reston, VA, 1986.

[10] Bylilnsky, Gene, "New Industrial Revolution Is on the Way." *Fortune,* October 5, 1981.

[11] Martin, John M., "CIM: What the Future Holds." *Manufacturing Engineering,* January 1988.

[12] Slattery, Thomas J., "Is CIM a Certainty?" *Machine and Tool Blue Book,* December 1985.

[13] Koves, Gabor, "Computer Integrated Manufacturing: An Overview." *Automotive Engineering,* June 1986, p. 85.

[14] Allen, David Chris, "Techniques: Computer-Integrated Manufacturing." *Datamation,* March 1980.

[15] Abraham, R. and Shrensker, W., "Computer-Integrated Manufacturing." *Computer-Integrated Manufacturing Series,* Society of Manufacturing Engineers, Dearborn, MI, 1986.

[16] CAM–I, "Computer Integrated Enterprise (CIE) Program." *Prospectus,* Arlington, TX, 1987.

[17] Merchant, M. Eugene, "The Automated Factory: Technology Paves the Way to Improve Productivity." *Compressed Air Magazine,* September 1983.

[18] Air Force Materials Laboratory, "U.S. Air Force Integrated Computer Aided Manufacturing." *ICAM PROGRAM PROSPECTUS,* Air Force System Command, Ohio, 1979.

[19] Sobczak, Thomas V. "A Glossary of Terms for Computer Integrated Manufacturing." Casa of SME, Dearborn, MI, 1984.

2

Computers in Manufacturing

Overview

Many interrelated manufacturing activities are grouped together to form a special application system that may be referred to as a production and control system (PACS). The grouping of manufacturing activities into PACS varies from one manufacturing environment to another. A PACS is defined as a subsystem in a global manufacturing environment. It may be a single subsystem, or it may be a complex set of subsystems. An illustration of PACS working in a global manufacturing system is shown in Figure 2.1. For PACS to meet their designed functional requirements, they should be designed to function independently of other PACS. Also, PACS should be able to work collectively with other PACS in a total integrated manufacturing environment. Each PACS in the total system can have an effect on the other PACS in the total system, and a systems planning approach must be taken for the following reasons:

- To prevent duplication of effort.
- To enable vital information to pass efficiently through the system.
- To allow each PACS to know its relation to the others and how it affects the others.
- To make the whole manufacturing system function more efficiently and productively [1, 109].

Computers are by far the most powerful single approach used in integrating and manipulating the series of interrelated manufacturing PACS and activities.

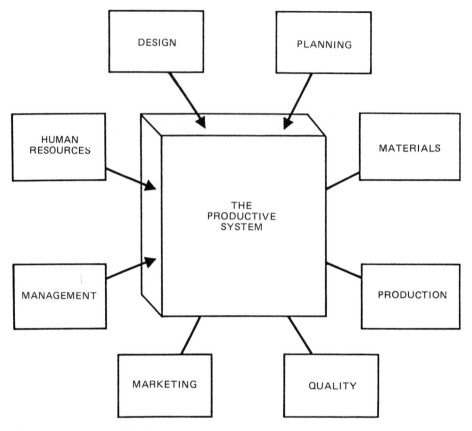

Figure 2.1 Interaction of PACS in a Manufacturing System

They have brought manufacturing technology into the era of "smart" machines. The advances in technical production have brought about a marriage between computer technology and manufacturing technology that has enhanced manufacturing technology development. This marriage is the basis for computer-aided production and control systems (CAPACS), which are computer-driven PACS. Thus, CAPACS have increased the roles of smart machines in production and control functions. The increased roles of smart machines have demanded a more intimate communication and interaction between such functions as design, financial and accounting, production, personnel, and marketing. The ways in which production operations are conceptualized, formalized, discharged, and performed are being changed by CAPACS.

Typical CAPACS in manufacturing are as follows:

- CAD   Computer-aided design
- CAIN   Computer-aided inspection

- CAM — Computer-aided manufacturing
- CAPP — Computer-aided process planning
- CAQC — Computer-aided quality control
- CIPM — Computer-integrated production management
- DNC — Direct numerical control
- GT — Group technology

Some of the foregoing terms (CAPACS) were discussed in Chapter 1. Others will be discussed in subsequent chapters.

Figure 2.2 gives an overview of interrelated functions of CAPACS working from an integrated data base system. The design data, generated by the interaction between CAPACS, is a single collection of all the information that describes the product and related operations. It is the hub of the manufacturing wheel. The CAD system is the principal tool used by engineering in carrying out its responsibility. The spokes of the wheel are made from various kinds of CAPACS involved in the

Figure 2.2 Interaction of CAPACS in a Manufacturing System

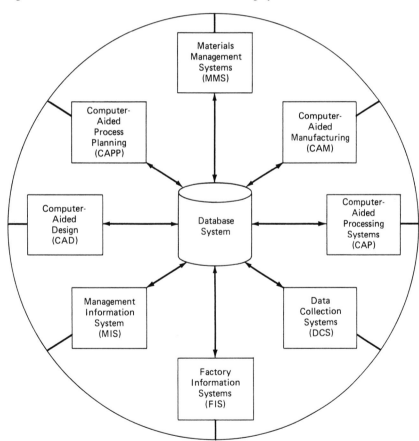

activity. Each CAPACS has a communication link to the controlled database so that it capture data it will use to form its own distributed database. Values are added to the distributed database to meet the needs and requirements of its expected users.

The application of CAPACS to the manufacturing process enables the total system to increase productivity, reduce waste, and produce things it would not otherwise be able to make. As a result, new technologies, demands for products of higher quality and lower production costs, and the need for improved technology in a competitive society have caused extensive use of CAPACS.

Automation Concepts

Automation may be defined as a system that is relatively self-operating. Such a system includes complex mechanical and electronic devices and computer-based systems that take the place of observation, effort, and decision by a human operator [2, 3]. It is a system that exhibits properties of human beings by following predetermined operations or responding to encoded instructions.

Computer Process Control

Process control involves the control of variables in a manufacturing process, where one or any combination of materials and equipment produces or modifies a product to make it more useful and hence more valuable. In process control systems, the computer serves as the control mechanism that automatically controls continuous operations. Two kinds of control systems are the *open loop* and the *closed loop*. In an open-loop control system, the computer does not itself automate the process. That is, there is no self-correction. The process remains under the direct control of human operators, who read from various sources of information such as instruments, set calibrated dials for process regulation, and change the controlling medium.

Closed-loop control systems use computers to automate the process. The computer is directly in charge of the process. It adjusts all controls from the information provided by sensing devices in the system to keep the process to the desired specifications, a technique that uses a feedback mechanism. Feedback is the action of measuring the difference between the actual result and the desired result (Figure 2.3) and using that difference to drive the actual result toward the desired result. The term *feedback* comes from a measured sample of the output of the process (production) function that becomes the input of the control function. That is, the output of the control function, meeting special designed requirements is the input to the control system. Thus, the signal begins at the output of the controlled production function and ends at the input to the production.

Typical functions of process control systems are monitoring, data logging, quality control, maximizing output, maximizing profit for a given output, supervisory control, and factory information systems (FIS). Benefits of computer process control systems are increased productivity, improved product quality, and increased efficiency, safety, comfort, and convenience.

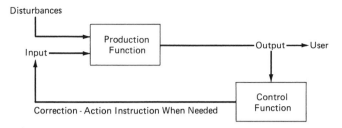

The Productive System . . .
Uses production operations and activities in such a manner as to transform a set of <u>input</u> parameters of new materials and shapes into a "designed" <u>output</u> of proper size configuration and performance specifications

Figure 2.3 A Block Diagram of a Closed-Loop Control System

Management Information Systems (MIS)

Management information systems are designed to aid in the performance of management functions. These systems are generated by computer systems and are developed to provide executives with up-to-the-minute information about the operations of the enterprise. When required, information systems are used to aid management in the decisionmaking functions of the enterprise. Viewing CIM as an information system for the enterprise for decision making, CAPACS must be information interconnected. As a result, there are many software packages associated with the CAPACS in Figure 2.2. Typical of these are CAPP, DCS, FIS, and CAD.

The concept of an MIS is a design objective, its goal being to get the correct information to the appropriate manager at the right time. As a result, MIS implementation varies considerably among manufacturing enterprises because of each organization's function, type of production, information resources available, and organizational commitment to MIS.

Engineering

Computers are used extensively in most engineering functions. Engineering is a profession in which a knowledge of the natural sciences is applied with judgment to develop ways of using the materials and forces of nature. Typical engineering functions using CAPACS are design, process planning, analysis and optimization, synthesis, evaluation and documentation, simulation, modeling, and quality control planning. Using CAPACS in engineering increases the productivity of engineers and improves the quality of designs.

For example, the application of computers to an engineering design process is performed by a CAD system (see Chapter 3). Engineers can design and thoroughly test concepts quickly and simply from one workstation. Computers permit engineers to take a concept from its original design through testing to numerical control (NC) output, or a combination of steps in between. They perform complex scientific and engineering computations rapidly with high accuracy; calculate physical properties before actual parts are made; and provide a fast, easy method to create models of even the most complex parts as shown in Figure 2.4.

The computer has influenced the way products are designed, documented, and released for production. As technology develops, engineering operations are becoming more and more automated and are relieving the engineer of many tedious manual calculations.

Production

Applications of computers to the production process (see Chapter 4) encompass such functions as computer monitoring, supervisory computer control, direct digital control (DDC) material handling, and production fabrication, assembly, and test/inspection operations. New ideas and technology developments are gaining acceptance on the factory floor. More important, the integration of more computers into the production process increases automation on the factory floor.

Computer automation helps to organize, access, and provide vital information in a common data base system for use by all manufacturing operations. Computer automation helps to control and to schedule machines and processes, and to route

Figure 2.4 A Geometric Model of a Part Generated by the Computer

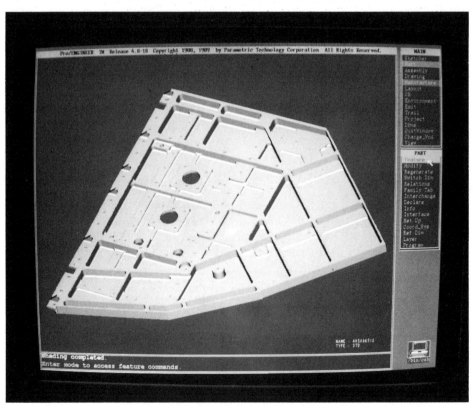

and to control raw materials and parts. A computer automated system concept is shown in Figure 2.5. Each function in manufacturing has its own area controller under the control of a host computer (such as the HP3000) in order to share information with other operations.

Computer-Aided Engineering and Production

The production cycle (CAD/CAM cycle) highlights four distinct phases in the manufacturing of a product, as shown in Figure 2.6. These phases are product definition, translation, construction, and support.

Definition is the engineering design process. *Translation* is the manufacturing engineering that provides the initiating actions for the manufacturing processes, including tool orders for the production tools to accomplish fabrication and assembly. *Construction* (production) consists of the physical actions of fabrication and assembly of the product. *Support* is the preparation of maintenance manuals, parts catalogs, and spares, together with field support of the customer. These basic phases are bound together by management control systems that provide scheduling, budgets, job tracking, shop loading and control, order writing, procurement and inventory control, and so on.

The use of computers in automating and integrating engineering and production systems (Chapter 5) adds a new dimension to the manufacturing process. A typ-

Figure 2.5 A Conceptual Computer Automated Factory

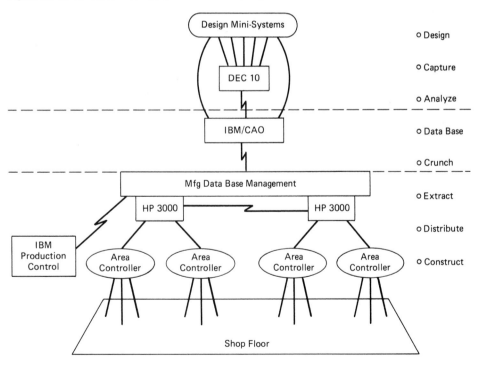

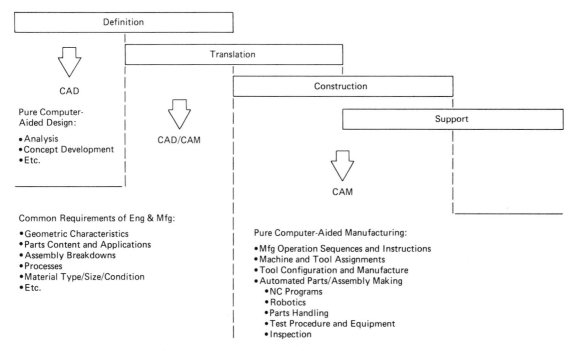

Figure 2.6 CAD/CAM Systems in Manufacturing

ical commercial integrated system is Control Data's "Integrated Computer-Aided Engineering and Manufacturing" (ICEM) system, which ties together the procedures required to take products from concept to production line. This system permits all functions involved—from design to manufacturing—to share data.

Integrated computer-aided engineering and production systems (CAD/CAM systems) streamline a firm's operation. They reduce design time, accurately create test conditions, and directly link them to production, all of which reduce production costs and time, thus making the price and availability of the end product more competitive. Most integrated CAD/CAM systems perform the following key manufacturing functions:

- Preliminary and detailed designs and drafting
- Solid geometric modeling
- Testing and analyzing models
- Finite element modeling
- Generating cutter-line output files for NC machines
- Process planning and group technology

Through computer-integrated CAD/CAM systems, factory automation, process control, design engineering, facilities engineering, and the design of material handling systems and quality control systems are integrated through a common data base system. As a result, the production cycle is shortened and better managed, and the time from product design to product shipment is greatly reduced.

Business

In today's manufacturing environment, computers are playing important roles in supporting various business functions. Typical of such key business functions in a manufacturing enterprise are production planning and control, finance and accounting, distribution management, maintenance scheduling and control, management information systems, data processing, and product planning. The applications of computers to assist or aid in performing various operations in these functions are referred to as computer-aided business (CAB). Chapter 6 discusses this topic in detail.

Computers are influencing changes in the way manufacturing enterprises carry out their business operations and manage production functions. Computers aid in planning for the accomplishment of objectives through effective management. They assist in planning and establishing exactly where, how, and when various activities that are part of a long-term program are carried out. They help planners produce optimized schedules, improve production line efficiency, and use manufacturing resource planning (MRP II) (see Chapter 6.), which is a formal system for planning and managing a production function's resources.

The use of computers has gone beyond the calculation of payroll or the fancy electric typewriter that can produce reports by the ton of paper. Computers are now contributing to the relief of many labor-intensive tasks and, with intricate programs, assisting the design process and the automation of the factory.

Computer applications are appearing in all areas of manufacturing operations. Each application has programming that satisfies the unique needs of that particular function or organization. Little attention, if any, has been given to the overall impact. The growth of isolated micro- and minicomputer installations in standalone environments has created a hodgepodge of information across the business operation. Yet each element has been properly cost justified and will indeed cause savings to accrue over previous methods. However, the resultant effect is one of unrelated empires, and short range planning is the treatment of manufacturing symptoms and not the disease. The term *islands of automation* is appropriate in such cases. The treatment of the disease is to work toward integrating the islands of automation.

Summary

Computers are changing the internal structure of manufacturing organizations, their methods of operations, and their external relationship to society. They assist in all manufacturing operations. A brief scenario of computers in manufacturing follows.

COMPUTERS IN MANUFACTURING

In this scenario, all action begins in engineering where new products are conceived that are presented to potential customers. The computer has a key role, and computer graphics is a core tool. It is normal practice in conceptual or preliminary design to make iteration after iteration to develop a product. Analysis and simulation methods also draw on the computer for support. A serviceable graphics software package provides these capabilities, or it interfaces to programs that will. When a concept is sold to a customer, program control develops master schedules and budgets for all functions of the project. Production design then puts the flesh on the skeletal concept, manufacturing engineering prepares the tool plan and the process plan, procurement orders material as defined on bills of material, tool design prepares designs to fulfill the manufacturing plan, tool manufacturing builds the tools, and production control (Figure 2.7) issues fabrication and assembly orders to make sure that the proper tools are in the right place at the right time and that a product is created. Quality assurance inspects parts and assemblies in conformance with the engineering definition of the product, and delivery to the customer is accomplished.

Central to the whole process is the engineering design. All functions relate to it for their actions. The ability of the designer to communicate concepts is of paramount importance. Interpretive issues are kept at an absolute minimum. Each change affects every function; even if no action is necessary, each change is analyzed to make that determination. A good computer graphics system that is user-oriented eliminates or at least minimizes changes. It also provides a communication medium for the designer that is far better than any method previously available. A policy of computer accuracy —that is, no out-of-scale dimensions with squiggly lines, no line breaks, and no uncontrolled changes—significantly reduces interference fits in assembly. The computer model provides visibility that helps the designer avoid such problems. The design file, locked but accessible to all downstream users, provides a single source of data to all

Figure 2.7 Control Systems in Manufacturing

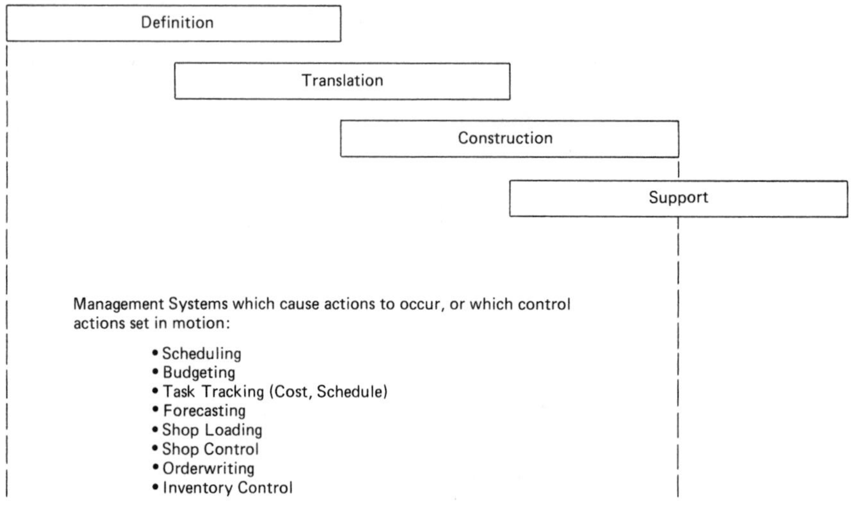

Management Systems which cause actions to occur, or which control actions set in motion:

- Scheduling
- Budgeting
- Task Tracking (Cost, Schedule)
- Forecasting
- Shop Loading
- Shop Control
- Orderwriting
- Inventory Control

functions in a read-only mode. This becomes the master drawing file, which no longer relies on vaulted vellums, and its proper protection is important. A daily redundant file spun off on tape should be vaulted, and weekly storage in an off-site location for disaster control must be provided.

Integration requires all the users of the engineering design, even engineering itself, to interface with the master drawing file for the extraction of data to satisfy their function in the chain of events leading to the end item. There are other tools that can assist in this process. Typical of such tools are NC processors, graphic systems, engineering/scientific processors, CAD/CAM workstations, and finite element solver.

Computer Control of Manufacturing Systems

Computer-controlled manufacturing systems use computers as an integral part of their control. As a result, computer controls are used in modern manufacturing automation from product inception, through product design, all operations between, and including product shipment and support. They control standalone systems such as robots, welding, spray painting, processing planning, and processing; they provide optimal control over the use of resources to produce a saleable product mix to satisfy sales forecasts and produce a profit for the firm; and they control complex systems such as automated storage/retrieval systems (AS/RS), automated guided vehicle system (AGVS), and flexible manufacturing systems (FMS). These concepts are discussed in later sections.

Computer-controlled systems are beginning to control many operations in the entire manufacturing cycle, running production lines, and taking over control of an entire factory. Even more challenging is the use of computer controls in data communication, database management, and the integration of the many islands of automation in the entire enterprise.

The factory of the future is built on the concept of an integrated control system (ICS), a new thrust in manufacturing that is beginning to mean a significant change in the very concepts of control.

Computer-Integrated Control System

The philosophy of an ICS is to tie together the subsystems of business, engineering, and production over the complete manufacturing cycle to create a smooth manufacturing business machine (CIM) [4, 2]. That is, an ICS accounts for all the variables in the process of doing business and for the interrelationships among these variables.

Integrated control systems are directed at changing the subtle differences in the meaning of control as the concept is used by the various functions in the manufacturing enterprise. They are also directed at tying together the management objectives for the firm and the various process functions required to produce the product. In summary, ICS are directed at creating and controlling a single, integrated plan for operating the facility and producing a cooperative, harmonious working relationship among all parts of the business.

Technologies such as computer systems, communication systems, programmable logical controllers (PLC), distributed control systems, and user-oriented information systems have been developed, perfected, and are being marketed for computer ICS. These systems are supported by perfected devices such as sensors, transducers, transmitters, receivers, measuring means, and communication media. Such components and systems are typical technological tools for computer-integrated control systems.

Levels of Control

Successful implementation of the factory of the future will be achieved by top-down design of a hierarchical control architecture as depicted in Figure 2.8 [4, 5]. Such a conceptual architecture should be implemented incrementally in a bottom-up fashion.

At the lowest level, Level 3, the system concentrates on control of individual machines or processes. This level receives directives from higher-level control systems or operators. It is the primary level of control and is usually the fastest-acting level.

At Level 2, related or sequential machines or processes are tied together dynamically. This level provides for interprocess coordination that produces optimum sequencing, throughput maximization, and improved use of capital and raw material. Simulators to predict upsets and bottlenecks or to aid engineers in

Figure 2.8 Hierarchical Automation System

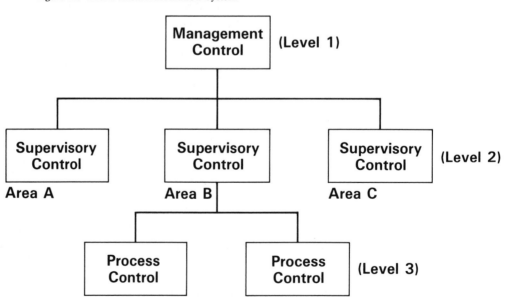

improving operations begin to appear. Usually, Level 2 controls take longer to respond to an action than at Level 3. In some cases, the control action is human-oriented.

The full system comes together at Level 1. Using resource-planning systems, high-level dynamic simulators, maintenance-planning systems and other tools, the entire complex can be put under control. This level of control is still intended for real-time control. However, control actions are slow, often in the order of days.

Architecture may be defined as any design or orderly arrangement perceived by man; that is, the present, physical, and logical arrangement of a computer. The architecture shown in Figure 2.6 is integrated through fast, reliable communication techniques and a single, well-defined database. A disadvantage of this architecture is that if any vertical line is broken, part of the total system is isolated.

Distributed Control

A review of Figure 2.8 shows that the control is distributed as shown in Figure 2.9. Thus, distributed control implies decentralization [5, 4]. The primary requirement of distributed control is the transfer of information from one station level controller to another, or from one computer to another on the same level. Thus,

Figure 2.9 A Distributed Control Concept

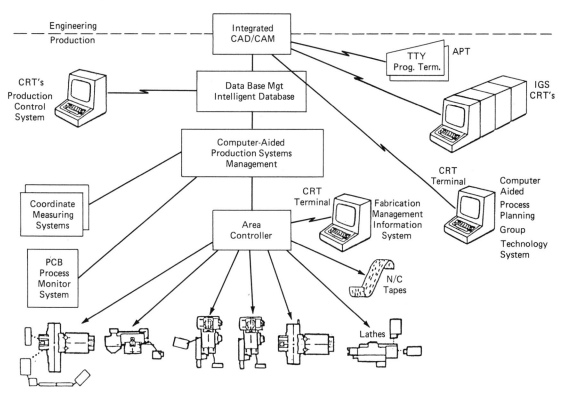

interlock information is passed among peer computers or controllers to coordinate machinery and process functions as the product moves from one step of the production process to the next.

Real-time control of machinery and industrial processes requires millisecond response to sensory input and coordination of related process elements. It is the real-time requirements that makes the decentralization concept so important. Supervisory and administration functions, where real-time communication is not so critical, must not interfere with the time-critical control loop.

Plant Communication

Communication is the transmission of intelligence between points of origin and reception without alteration of sequence or structure of the information. It includes data communication with its unique syntax. Data communication generally relates to the movement of computer-encoded information by means of electrical transmission. The heart of a factory communication system is a digital computer or group of interconnected computers, each of which may have human, database, machine, and terminal interfaces. These devices range from the simple to the intelligent. They are linked to a host computer and a complete network of terminals, concentrators, remote processors, machines, modems, multiplexers, and so on to form a complex system.

Each computer is selected on the basis of its ability to perform the task required at its location. The individual processors vary considerably in their speed, interfaces, languages, and other characteristics. However, it is important only that they be efficient in the performance of their assigned responsibility. As a result, a plant communication system is a complicated integrated system that must transmit data in both directions between a central location (host computer) and remote terminals, computers, databases, machines, and so on. Under the control of a computer, the system must get reliable and quality data to the user at the right time. A key communication issue is access to the up-to-the-minute data in the control systems of the plant floor and all other operations to eliminate costly errors, to improve product quality, and to improve productivity.

Network Architectures

The term *architecture* is commonly used today to describe networks. A network architecture, such as illustrated in Figure 2.10, describes the components in the network, how they operate, and what form they take. A network encompasses hardware, software, data link controls (DLC), standards, topologies, and protocols.

Hardware

The physical equipment in the network is referred to as the hardware. Typical of such equipment are computers, modems, telephones, terminals, machines, controllers, and the like.

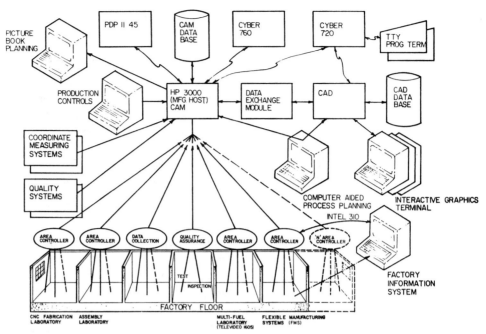

Figure 2.10 A Typical System Architecture Model

Software

Software consists of a set of programs, procedures, and sometimes associated documentation of the operation of the communication system. Communications software (CSW), applications programs (AP), network control programs (NCP), operating systems (OS), and database management systems (DBMS) are examples of software.

Protocols

A protocol defines how network components establish communications, exchange data, and terminate communications. A protocol is essentially a formal set of conventions, or rules, between communicating processes about the format and content of messages to be exchanged. Handshaking and line discipline are also protocols. To make implementation and usage more convenient in sophisticated networks, higher-level protocols may use lower-level protocols in a layered fashion.

Data Link Controls

A data link consists of electronic equipment that permits automatic transmission of information in digital form. It is an assembly, or installation, of equipment such as electronic devices, terminals, and machines and the interfacing circuits that are operating in a particular way to permit information to be exchanged between installations. The specific method of operation is defined by transmission codes, transmission modes, and direction control.

There are two major categories of link control protocol in use today:

1. Asynchronous, in which bit streams are transferred and received at nonuniform (asynchronous) rates.
2. Synchronous, in which bit streams are transferred at fixed rates with the transmitter and receiver operating in synchronization by the clock.

The flow of data to and from the many points and links within the network must be controlled and orderly. The sending and receiving sites must know the identification and sequencing of the messages being transmitted among all users. The connection path between sites is usually shared by more than one user, as in a multipoint configuration. A multipoint line, sometimes called a multidrop line, is a communications line having several subsidiary controllers that share time on the line under the control of a central site.

Data link controls provide for all these needs. They manage the flow of data messages across the communications path, or links. In essence, they are special kinds of protocols consisting of a combination of software and hardware located at each site in the network. Through a communication link, the DLC protocol provides for the reliable interchange of information between data terminal equipment. Typical DLC functions in the network are the following:

- Synchronizing the sender and receiver.
- Controlling the sending and receiving of data.
- Detecting and recovering transmission errors between two points.
- Maintaining awareness of link conditions.

Topologies

Topology describes the surface layout of the elements comprising the communications network—that is, it is the shape of the system. Communication systems may be implemented in a variety of ways. Typical communication systems are telephone lines, radio wave links, networks, serial connections, parallel connections, and power line system modulations. The choice of topology depends on factors such as cost, response time, throughput, capacity, load sharing, and capacity needs.

Standards

Systems for data transmission are available from many vendors who provide products for specific services and operations. Such a heterogeneous group of systems causes many problems in integrating the many islands of factory automation. Also adding to the problems of integration are the following:

- Lack of universal interface, communication, control, and protocol standards among makers of the various systems.
- Varying existing standards because of development by many standards organizations.

- Lack of agreement on a universal set of standards among industries, organizations, manufacturers, and users.

Unless a predetermined set of standard parameters is specified for the operation of systems in the factory of the future, trying to connect equipment from various vendors will be chaotic. Furthermore, these problems seem to be impeding the implementation of such factories.

Several standards, however, are accepted worldwide. They were developed by government agencies, associations, organizations, and industries working individually and sometimes collectively. Typical of such standards are the following

Standard	Purpose
RS232	Provide interface between the data communications equipment (DCE) and data terminal equipment (DTE). Such a connection is shown in Figure 2.11.
X.25	Interface between DCE and DTE for terminals operating in the packet mode on public data networks.
ASCII	American Standard Code for Information Interchange. A data communication code set for serial transmission.
IEEE-488	A general-purpose interface bus (GPIB) allowing two-way parallel communication between the computer and one, or more, external devices.
SDLC/HDLC	Synchronous Data Link Control (SDLC) and High-Level Data Link Control (HDLC). Protocols are defined to control the logic flow of information between two or more computers. They define the method by which the remote computer stations are addressed and how the stations communicate with each other.
IEEE-583 CAMAC	A nuclear instrumentation system interface system. The standard defines the mechanical configuration, the electrical connectors, the data transmission paths, and the protocol [3, 290].

Local Area Networks

A local area network (LAN) is a communications network operating in a local area. It extends from several hundred to several thousand feet within a building or other facility. A LAN is owned and operated by an individual or organization and is, therefore, not subject to regulation by either the Federal Communications Commission (FCC) or the state Public Utility Commission (PUC). A LAN is a means of connecting various types of equipment for the purpose of sharing resources and communicating in a distributed processing environment, as illustrated in Figure 2.12 [7]. That is, all devices that must communicate with one another in a CIM environment can be tied together through some form of common interface such as a LAN.

I/O Pin #	LEAD COLOR	SIGNAL NAME
1	Black	Protective ground (P GND)
2	Red	RS-232C Serial Output (S OUT)
3	Orange	RS-232C Serial Input (S IN)
4	Black-white	Request to send (RTS) Output (Printer Busy)
5	Black-green	Clear to send (CTS) Input*
6	Black-blue	Data set ready (DSR)*
7	White-red	Signal ground (SGND)
8	White-blue	Received Line Signal Detect (RLSD)*
9	Not Used	
10	Not Used	
11	Not Used	
12	Not Used	
13	Not Used	
14	White	Current loop output − (OUT−)
15	Green	Not used
16	Blue	Current loop output + (OUT+)
17	Black-red	Not used
18	Black-orange	Current loop input − (IN −)
19	White-black	Current loop input + (IN +)
20	White-green	Data terminal ready (DTR) output
21	Not Used	
22	Not Used	
23	Not Used	
24	Not Used	
25	Not Used	

Figure 2.11 RS232 Technique of Interfacing

An effective LAN has specific characteristics. Typical of these are the following:

- Is in a moderate-sized geographic area (office, laboratory, building, factory, industrial complex, or campus, for example).
- Has medium- to high-speed data channel(s).
- Provides reliable communications.
- Is owned and used by a single organization.
- Connects independent devices rather than a device and its peripherals.

A LAN usually has four major components that serve to transport data between end users: user workstation, protocol control logic, medium interface, and physical path.

User Workstation

The user workstation is used to accomplish an application. The station can be anything from a word processor to a mainframe computer.

Figure 2.12 A Typical Network

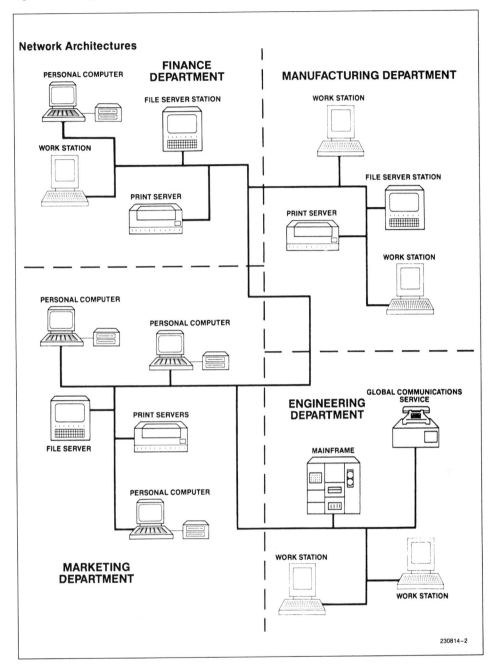

Protocol Control Logic

A protocol control logic takes the user's information and converts it to a protocol that can move over the LAN network to reach the desired location. The protocol control logic also controls the LAN and provides for the end user's access to the network.

Medium Interface

The medium interface function generates the electrical signals to be transmitted on the LAN. The interface is between the path and the protocol control logic and can take several forms, typical of which are CATV (cable television) taps, infrared diodes for infrared paths, microwave antennas, or complex laser-emitting semiconductors for optic fibers.

Physical Path

A LAN's path may consist of coaxial TV cables, coaxial baseband cables, twisted pair of wires, optic fibers, and microwaves. Cable TV coaxial cable is used in many networks because it has a high capacity, a very good signal-to-noise (S/N) ratio, low signal radiation, and low error rates.

Manufacturing Automation Protocol

There are many types and sizes of networks. Some LAN systems are owned by an organization and are confined to the premises of that organization. Others are owned by national or international public networks that may be accessed by the user. In many cases, LAN systems may also connect to public networks and be able to access them.

Networks are also connected in several configurations (typologies) (Figure 2.13) [7]. Each topology can constitute an entire network or can be just a portion of a large network.

A *star network* is a centralized network in which each device is connected individually to a central controlling point.

A *ring network,* sometimes called a loop, is a network in which each device is connected to the next in sequence in a closed, circular fashion. This topology has two possible paths, one in each direction.

A *bus network* is a topology that consists of a single shared line to which all devices are connected.

In a *hierarchy network,* a central root node has control over the entire network. The secondary nodes attached to the root can be front end processors (FEP) that, in turn, connect to tertiary nodes of lesser significance. The hierarchy can consist of a large computer that has overall control of several smaller control devices connected to their own terminals.

A *multiply connected topology* requires that there be at least two paths between any pair of nodes to assure availability and reliability in the event of failure of any one path or node.

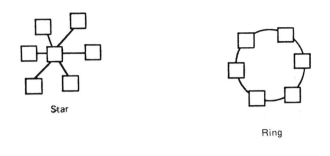

Star

Ring

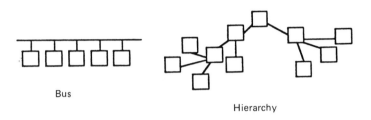

Bus

Hierarchy

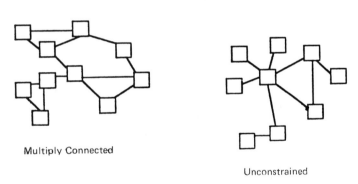

Multiply Connected

Unconstrained

Figure 2.13 Typical Network Topologies

The *unconstrained* network occurs most commonly. The term *unconstrained* merely means that the configuration is not restricted to one specific kind of connection.

Many LAN systems are available today. All use either basebands, broadbands, or both. In a baseband, the line is pulsed, as in the on/off conditions. Because these pulses are still the square wave direct current, they require no additional special handling or modulation, which makes the process relatively inexpensive. It is impossible to mix voice and data unless the analog voice signals have been previously digitized.

The broadband technology requires a conversion process and is therefore more expensive. The advantages gained are that voice, video, and data can all be sent on the same channels. Facilities can be shared through assignment of frequency channels to allow all devices to operate simultaneously. Popular modulation and multiplexing techniques can be used because a wide range of frequencies is available.

Typical vendor offerings of LAN systems are the following:

COMPANY: Digital Equipment
NAME: Decdataway
PATH & TYPE: Twisted-pair wire; bus

COMPANY: IBM (GS)
NAME: Series 1/Ring
PATH & TYPE: Coax; ring

COMPANY: Prime
NAME: PRIME NET
PATH & TYPE: Coax; ring

COMPANY: Xerox Corp
NAME: EtherNet
PATH & TYPE: Baseband coax; bus

COMPANY: AMDAX Corp.
NAME: Cable Net
PATH & TYPE: Baseband coax; bus

Open System Interconnect

In an effort to encourage open networks, the International Standards Organization (ISO) has developed the open system interconnect (OSI) reference model illustrated in Figure 2.14 [7]. In simple terms, the model logically groups the functions and sets of protocols necessary to establish and conduct communication between two parties. The OSI model consists of seven functions, often referred to as layers. It describes the functions of each layer in broad terms, not specific implementations.

Implementations

The layer function of the OSI model is summarized in Figure 2.14. Additional layer functions are discussed in the following several paragraphs.

Physical layer. The physical layer (1) describes the physical medium over which the bit stream is to be transmitted. This layer specifies type of cable (coax, twisted pair, and so on), signal levels, bit rate, data-encoding method, modulation method, and the method for detecting collisions in contention networks. Contention networks operate on a first come, first served basis. There is no initiation

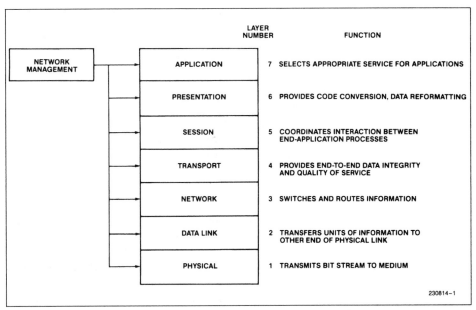

	LAYER NUMBER	FUNCTION
APPLICATION	7	SELECTS APPROPRIATE SERVICE FOR APPLICATIONS
PRESENTATION	6	PROVIDES CODE CONVERSION, DATA REFORMATTING
SESSION	5	COORDINATES INTERACTION BETWEEN END-APPLICATION PROCESSES
TRANSPORT	4	PROVIDES END-TO-END DATA INTEGRITY AND QUALITY OF SERVICE
NETWORK	3	SWITCHES AND ROUTES INFORMATION
DATA LINK	2	TRANSFERS UNITS OF INFORMATION TO OTHER END OF PHYSICAL LINK
PHYSICAL	1	TRANSMITS BIT STREAM TO MEDIUM

NETWORK MANAGEMENT

230814-1

Figure 2.14 Open System Interconnect Model (The International Standards Organization)

of communications from the central unit. Dial-up systems are very good at contention.

Data link layer. The data link layer (2) describes the rules for transmitting on the channel (made up of the encoder/decoder, transceiver cable, and transmission medium). Such items as the format of the information (frame) and procedures for gaining control of the channel (access method), transmitting the frame, and releasing the physical medium are specified by the data link layer.

Network layer. The network layer (3) controls switching between links in a multidrop network. A multidrop line has two or more service points connected to a single line; that is, it is possible to have several transmissions connected to the same transmission line. The network layer is not necessary for a single LAN because all stations connected to a LAN share the same channel. This layer is critical in gateway, communication serve, and dial-up applications.

Transport layer. The transport layer (4) ensures end-to-end message integrity and provides for the required quality of service for exchanged information. For example, end-to-end acknowledgments and flow control are performed by the transport layer.

Session layer. The session layer (5) establishes and terminates logical connections between network entities, and it is also responsible for the mapping of logical names into network addresses.

Presentation layer. The presentation layer (6) provides for any necessary translation, format conversion, or code conversion to put the information into a recognizable form.

Application layer. The application layer (7) provides network-based services to the end user. Examples of network services are distributed databases and electronic mail. The application layer should not be confused with the end user application itself.

Network management is responsible for operation planning, which includes the gathering of operational statistics such as errors and traffic. It is also responsible for network initialization and maintenance (fault isolation). Network management interfaces to each layer.

The OSI model has two key advantages. First, layers allow a clean division of the design task, making specifications clean. Second, systems based on a layered architecture are flexible because each level functions independently of the layer preceding or following it. Thus, specific layer implementations can be changed easily. For example, layers 1 and 3 of a network can be changed to be based on either CSMA/CD (IEEE 802.3) or token ring (IEEEE 802.5) without affecting layers 3 through 7.

A carrier sense multiple access/collision detect (CSMA/CD) protocol is a method of transmitting information in a LAN environment where only one transmitter may be on the line at any one time. CSMA/CD protocols use a nonpersistent carrier sense where stations transmit with a random delay to avoid repeated collisions. In nonpersistent, the user senses the channel and transmits the message if it is not busy.

A token ring protocol, sometimes called empty slot protocol, is usually on a ring topology. In such a case, a token is a time slot, or packet, that is passed to the next station on the ring network. The packet may contain the address of the station or may simply be an empty slot available to any station that has traffic to place in the packet.

Manufacturing Automation Protocol

The manufacturing automation protocol (MAP) is a broadband, token-bus-based seven-layer communications standard designed to integrate the islands of factory automation. It is a multivendor LAN defacto standard for factory networking organized by General Motors. To date, most of the big players in factory automation have endorsed it. The MAP incorporates international standards from the Institute of Electrical and Electronics Engineers (IEEE), the National Bureau of Standards (NBS), and the Open System Interconnect (OSI) models of the International Standards Organization (ISO).

The Map Model

A generic model of a MAP is shown in Figure 2.15. A comparison of this model with the one shown in Figure 2.14 shows that the generic model is based on

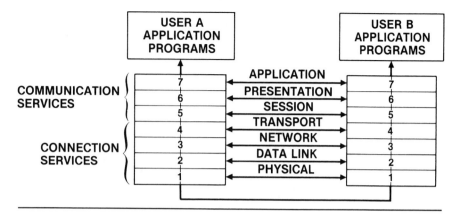

OSI MODEL

MAP

- SEVEN LAYER
- BROADBAND
- TOKEN BUS
- STANDARD LIMITER
- EVOLVING

Figure 2.15 A Generic Model of MAP—the Seven-Layer Model

the OSI model. However, the MAP model is well defined around proven standards. The need for standards used in the MAP can best be seen as part of the more fundamental problems of underuse of materials and equipment in today's automated but chaotic factory. A MAP appears to provide the key to a practical CIM facility— communication. Communication is the key to getting the most out of manufacturing equipment.

A factory MAP logical hierarchy is illustrated in Figure 2.16. Such a generic concept of a hierarchy gives factory information management the ability to provide information where it is needed when it is needed. It provides information flow in both vertical and horizontal directions to establish peer-to-peer communications.

Figure 2.17 illustrates a diagram of a comparison of the Allen-Bradley approach to the communication problem based on a MAP [5]. This technique enhances each island of automation and brings genuine automated status to a modern manufacturing facility. Figures 2.18A and B show an OSI seven-layer model connection between two users. Figure 2.18A provides for connection between two users, whereas Figure 2.18B establishes connection between two users through a gateway, which is a connection for two networks. The gateway can be a connection between a local network and a packet, or satellite, network. It also can connect a network of terminals to a host computer.

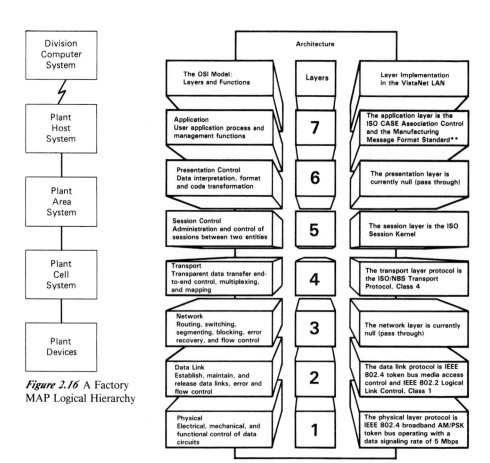

Figure 2.16 A Factory MAP Logical Hierarchy

Figure 2.17 ISO and VistaNet Networking Protocol Structures

Figure 2.18A Connection Established Between Two Users

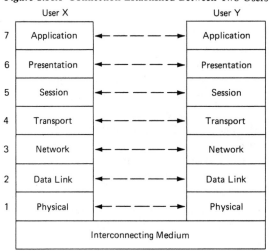

Fundamentals of Product Processes and Operations

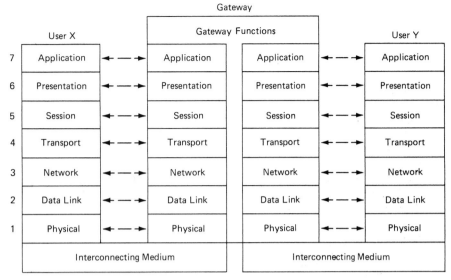

Figure 2.18B OSI Gateway Connections

Typical benefits of a MAP are the following [10]:

1. ***Flexibility***

   Transportability of applications (same network services available to all applications).

   To network or develop custom solution, user installs best equipment for the application instead of a compromise.

   Cabling chosen because it allows equipment relocation without rewiring (this provides a lower cost).

2. ***Functionality***

   Eliminates Islands of automation.

   Multivendor environment is realistic.

   Standard messaging for factory floor device.

   More direct access to information.

   Common network "view" for the user, minimizes training and confusion (doesn't have to know vendor specifics—for example, VTP solution).

3. ***Implementability***

   Cuts implementation time (no need to engineer the connectivity solution).

   Allows user to focus at a higher level above the network because the MAP communications solution is a given.

4. ***Low Cost of Ownership***

   Reduces cabling costs in new installations.

   Reduces rewiring cost or equipment rearrangement.

Less training involved; more comprehension of communications because there is only one system to learn.

Leveraging applications (can be installed at many sites even if different vendors' equipment is used because the communications interface is the same). Volume production of Very Large-Scale Integrated Circuit (VLSI) should lower overall equipment costs. An Integrated Circuit (IC) is an electronics device containing several elements, active or passive, that performs all or part of a circuit function.

5. *Maintainability*

Common set of tools across multivendor systems.

"Best case" set of tools can be developed on basis of focus through use of standards.

Fewer people required to maintain the communication system.

6. *Predictability*

Products will be certified using NBS conformance tools.

Users will know before purchase that product performs up to the standard.

Eliminates uncertainty of debugging vendors' products.

7. *Reliability*

More direct access eliminates points of failure.

Cabling more reliable.

VLSI provides more inherent reliability.

8. *Simplicity*

Connectivity simplicity (fewer active parts in the system).

Network management simplicity (because of tools available even though networks will be larger and more complex in configuration).

Manufacturing Databases

Modern manufacturing is an extremely sophisticated and complex activity. Much data processing and control are involved. One continuing factor in this vast data operation is the great deal of automation that has been infused into the factory environment during the 1980s. The complexity of manufacturing systems and the amount of data required for efficient operations are expected to continue growing. As a result, computer systems are needed to operate the automated systems and manage the data. The development of such systems is a formidable challenge.

Manufacturing data are stored in databases. A computer database is a collection of electronic information (data) records. The data are carefully stored in a specific, regular way in accordance with rules defined by a particular database management system (DBMS), as discussed in Chapters 5 and 7. With the aid of a computer, a DBMS can create a database and access, store, retrieve, and manipulate the data.

Manufacturing databases of various kinds are in use to meet the specific requirements of many diverse users. A manufacturing database consists of files of data that contain two distinct types of information: operational data and association data. Operational data include such information as part numbers and tool types, whereas association data describe the relationship between information in the database, indicating, for example, which machine performs which operation on a part.

Operational data are typically stored in individual records consisting of fields of data. A field is a single data item or a subdivision of a record. A record might, for example, contain the part number, part name, and weight for a specific part in three fields. A collection of similar records is stored in a file, which is a group of records. All the files comprise the database. Uses of data files in manufacturing are shown in Figure 2.19. The factory of the future will require an integrated database system. An integrated database system will share data with the various users under controlled conditions.

Database Technology

Advances in computer science technology provide the designer of software systems with powerful operating systems, high-level computer languages, structured programming concepts, and database technology. Computer science technology permits software designers to face database challenges with tools and concepts that were unavailable only a few years ago.

Figure 2.19 Applications of Data Files in Manufacturing

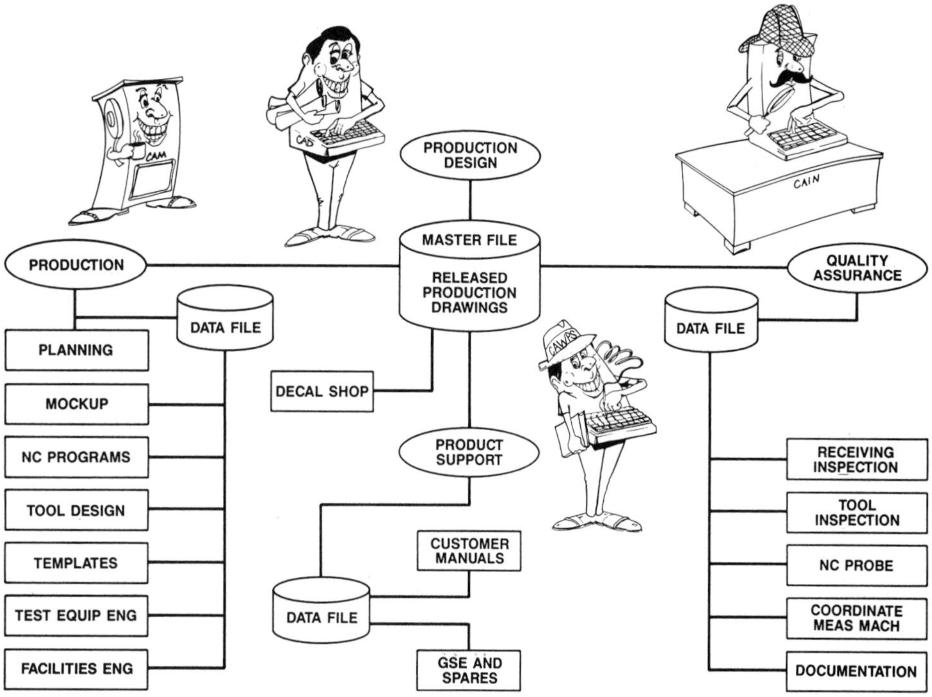

The use of database technology involves a switch in mind-set for many designers when they develop software for a factory environment. During the early development of software, most applications were primarily of a business and financial nature. Very few applications during the early stages were specifically related to factory automation in the DBMS.

Databases for the Factory of the Future

The database for the factory of the future may resemble the database illustrated in Figure 2.20A [11]. This model contains all relevant data generated by various functions (Figure 1.6) combined in a single geometric database. Various functions can readily extract data from the database to further their operations. A typical example of data retrieval is in the operation of a flexible manufacturing system (FMS) as discussed in Chapter 5.

An FMS incorporates many islands of automation concepts into a single production. Computers play important roles in the concept of islands of integration as well as in an individual automation island.

Programmable Logic Controller

Programmable logic controllers (PLC) are used extensively on the factory floor. A PLC is based on solid-state digital logic and built of computer subsystems. It is primarily intended to take the place of electromechanical relay panels in applica-

Figure 2.20A A Conceptual Product Model

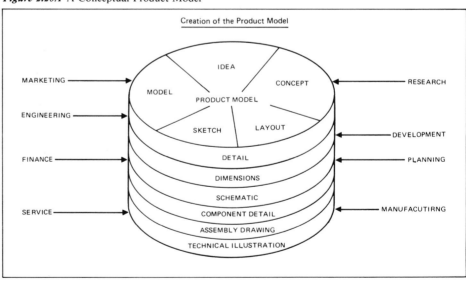

INTEGRATED ENGINEERING AND MANUFACTURING ENVIRONMENT

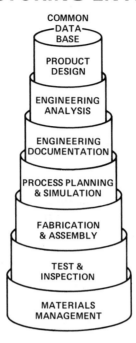

Figure 2.20B Distributed Database Concept

tions in which rewiring is made necessary by periodic changes in sequence. This type of controller is particularly useful in the control of processes, materials handling, and certain machine functions.

Exercises

Homework Assignment

1. Identify several manufacturing areas in which computers are used.

2. Explain the differences, if any, between CAPACS and control systems.

3. Explain the differences, if any, between CAPACS and automation.

4. a. Identify areas in business, engineering, production, and human resource functions where computers are used.

   b. Explain how these computer applications can improve productivity, reduce cost in manufacturing, and improve product quality.

5. a. Define the terms listed in Item 5B.

   b. Identify the major manufacturing function in which each of the following is a subsystem (subset):

1. CAD	5. CAQC
2. CAIN	6. CIPM
3. CAM	7. GT
4. CAPP	

6. Define a LAN and explain its role in the factory of the future.

7. Explain how computers are changing the internal structure of manufacturing organizations, their methods of operations, and their external relationship to society.

8. Explain why some manufacturing managers are somewhat reluctant to make more use of computers in manufacturing applications. Discuss how you may influence them to use more computers.

9. Define protocol, and list several major functions.

10. a. Define the seven-layer OSI model and its purpose.

    b. Which layers of the OSI model are physically connected? Which layer contains four DLC protocols?

11. a. Describe the difference between asynchronous and synchronous protocols.

    b. List some advantages and disadvantages of each type of protocol.

12. List some well-known DLC protocols.

13. Explain the difference between a closed-loop and open-loop control system.

14. a. Name the four distinct phases in the manufacturing of a product.

    b. What is the main function of each?

15. a. Name several types of software used in a CIM facility.

    b. Give a function of each.

16. What are the differences, if any, among the ISO, the seven-layer OSI Model, and a MAP?

17. Define:

    1. Database
    2. DBMS
    3. Record
    4. Field

18. Explain a major difference between a PC and a PLC.

1. Interrelated manufacturing activities are grouped together to form special applications systems called
   1. Production and control systems
   2. Computer-aided production and control systems
   3. PACS
   4. CAPACS
   5. All these.

2. The application of CAPACS in manufacturing enables the system to:
   1. Increase productivity
   2. Reduce waste
   3. Produce complex parts
   4. All these.
   5. None of these.

3. Integrated engineering and production systems integrate the procedures required to take products from concept to production lines.
   True   False

4. Computer-controlled systems have begun to operate over the entire manufacturing cycle. Typical of such operations is (are):
   1. Data communication
   2. Data base management
   3. Running production lines
   4. All these.
   5. None of these.

5. A physical path of a LAN may consist of:
   1. Coaxial TV cable
   2. Optic fibers
   3. Microwaves
   4. All these.
   5. None of these.

6. A _____ describes the surface layout of the elements comprising the communication network.
   1. Topology
   2. Standard
   3. Network interface
   4. All these.
   5. None of these.

7. A LAN in most manufacturing facilities is subjected to regulations by either the FCC or the PUC.

True   False

8. Distributed computer control implies computer decentralization.

True   False

9. Automation is the key to:

1. A shorter work week

2. Safer working conditions for the worker

3. Lower prices and better product

4. Means of increasing our standard of living

5. All these.

10. In digital data transmission, "handshaking" or the "please-and-thank-you" routine is associated with

1. Synchronous transmission

2. Asynchronous transmission

3. Conversion transfer

4. Items 1 and 2.

5. All these.

11. Protocol is the characteristic of a communication interface that defines the:

1. Format of the data

2. Meaning of the control signals

3. Order in which messages are transmitted

4. Method used to check for errors in transmissions

5. All these.

12. A standardized unique seven-bit binary code used extensively in serial data transmission is the

1. CAMAC (or IEEE–583) interface system

2. ASCII interface system

3. IEEE–488 interface system

4. SDLC/HDLC interface system

5. All these.

13. A major concern in the use of microcomputers in control systems is the need for communication among the various parts of the system.

True   False

14. The _____ interface system was originally designed for nuclear instrumentation labs where instruments were constantly swapped among systems.

    1. CAMAC (or IEEE–583)

    2. ASCII

    3. IEEE–488

    4. SDLC/HDLC

    5. None of these.

15. A characteristic of a digital communication interface is that data may be transmitted synchronously or asynchronously in either serial or parallel manner.
    True   False

16. A flexible manufacturing system (FMS) consists of a group of processing stations (usually NC machines) connected by an automated workpart handling system.
    True   False

17. A half-duplex channel can transmit data in only one direction.
    True   False

18. Asynchronous transmission is used for high-speed transmission of a block of characters.
    True   False

19. A distributed database is one set of data that belong together logically and that are stored at two or more different physical locations.
    True   False

20. A _____ cable consists of one conductor, usually a small copper tube or wire, within and insulated from another conductor of large diameter, usually copper tubing or copper braid.

    1. Coaxial cable

    2. Optic fiber

    3. Twisted-wire pair

    4. Items 1 and 3.

    5. All these.

21. _____ data processing provides computer capability on multiple processors that are capable of processing independently and that also interact regularly by exchanging data.

    1. Interactive

    2. Distributed

    3. File

    4. Real-time

    5. Batch

22. Engineering is the principal user of CAD, whereas the factory floor is the principal user of CAM.
    True   False

23. The computer serves as the control mechanism and manages the process operation.
    True   False

24. The translation phase of product manufacturing is under the responsibility of the engineering design process.
    True   False

25. Computers are changing the internal structure of manufacturing organizations and their methods of operations but with small changes in product support.
    True   False

26. Successful implementation of the FOF will be better achieved by bottom-up design of a hierarchical control architecture.
    True   False

ANSWERS TO SELF-STUDY TEST

1. 5	**7.** False	**13.** True	**19.** True	**25.** False
2. 4	**8.** True	**14.** 1	**20.** 1	**26.** False
3. True	**9.** 5	**15.** False	**21.** 2	
4. 4	**10.** 1	**16.** True	**22.** True	
5. 4	**11.** 5	**17.** False	**23.** True	
6. 1	**12.** 2	**18.** False	**24.** False	

References

[1] Davis, G. B., "CAM: A Key to Improving Productivity." *Modern Machine Shop,* September 1980.

[2] Groover, Mikell P., *Automation, Production Systems, and Computer-Aided Manufacturing.* Prentice-Hall, Englewood Cliffs, NJ, 1980.

[3] Bateson, R., *Introduction to Control System Technology,* 3rd ed. Merrill, Columbus, OH, 1988.

[4] Pujol, J. B., *Integrated Control Systems: Why and How.* Rust International Corporation, Birmingham, 1983.

[5] Allen-Bradley, "Industrial Automation Productivity Issues." *Plant-Wide Communication,* Cleveland, OH, 1984.

[6] Black, U. D., *Data Communications and Distributed Networks,* 2nd ed. Prentice-Hall, Englewood Cliffs, NJ, 1987.

[7] INTEL, *LAN Components User's Manual.* Intel Corporation, Santa Clara, CA, 1984.

[8] Appleton, D. S., "The CIM Database." *Computer-Integrated Manufacturing Series,* Vol. 1, No. 4. Computer and Automated Systems Association of SME, Dearborn, MI, 1986.

[9] Chiantella, N. A., "Achieving Integrated Automation Through Computer Networks," *Computer Integrated Manufacturing Series,* Vol. 1, No. 2. Computer and Automated Systems Association of SME, Dearborn, MI, 1986.

[10] Yeager, G., "Understanding MAP." *Integrated Automation Corporation,* Ypsilanti, MI, CASA/SME, 1985.

[11] Weber, T. J., "The CAD/CAM Data Base . . . The Foundation of CIM." *Commline,* January–February 1985.

[12] Kinnucan, Paul, "Computer-Aided Manufacturing Aims For Integration." *High Technology,* May–June 1986.

[13] Drozda, T. J., Stranahan, J. D., and Farr, G., *Flexible Manufacturing Systems,* 2nd ed. SME, Dearborn, MI, 1988.

[14] Koren, Yoram, *Computer Control of Manufacturing Systems.* McGraw-Hill, New York, 1983.

[15] Sobczak, J. V., "A Glossary of Terms for Computer Integrated Manufacturing." CASA/SME, Dearborn, MI, 1984.

3

Computer-Aided Engineering Systems

Overview

Engineering plays an important role in the manufacturing process. Computer-aided engineering (CAE) includes design, analysis, process planning, numerical control programming, mold and tool design, quality planning, and product support services to many functions in manufacturing. Computer-aided engineering is active throughout the manufacturing cycle (Figure 1.6) from the definition of product requirements through product support. These functions are defined and discussed in other sections of this text.

During the manufacturing process, CAE is applied. In a modern manufacturing enterprise, the primary application and activity flow of each system is both implemented and monitored by computer-based subsystems covering such areas as design, analysis, process planning, numerical control, mold and tool design, and quality control of production. Thus, the application of computers to aid in the engineering processes is referred to as computer-aided engineering (CAE). As a result, many CAE subsystems are tightly coupled under the umbrella of CAE. Some of the basic subsets of computer-aided engineering are discussed in this chapter.

In CIM, another way of doing manufacturing business, computers integrate product requirement data and operations as previously shown in Figure 1.7. It provides timely, accurate, and reliable data to these operations when they request it. CIM also promises dramatically improved capacity for producing product variety with greater opportunity for design based on customer specifications. Functional design strategies can facilitate the design of customer-initiated variety in a make-to-order situation, a feature which should stimulate demand for CIM-based production in general.

Customer-specified variations in design will be possible with an improved human interface to the CAD process. This interface will require an intelligent system which can interactively formulate functional requirements on the basis of customers' requests. The functional requirements will then act as constraints on the CAD process, greatly reducing the search space of design possibility.

Computer-Aided Design

Introduction

The use of computers in engineering has radically changed the design process as well as the resulting designs. Using techniques such as solid modeling, finite-element analysis, and dynamic simulation, the designer can design the strength of the structure to match precisely the expected operating loads. Design engineers can now develop complex and innovative designs that would have been impractical not many years ago.

Computer-aided design (CAD) may be defined as the use of computer systems to assist in the creation, modification analysis, or optimization of a design. Interactive computer graphics (ICG) technology is the foundation of CAD. It permits the user to communicate with a graphic system using input devices such as a mouse, digitizer tablets, joy stick, or light pen. The computer communicates with the user via a cathode ray tube (CRT) or hard copy devices such as a printer or plotter. The various design-related tasks that can be performed by a CAD system may be grouped into three functional areas:

1. *Geometric Modeling.* The CAD system constructs the graphic image of a part on a CRT by using basic geometric entities such as points, lines, and circles. The graphic image can then be manipulated by the designer by using the commands of the software. Typical of these manipulations are scaling, transformation, and rotation. As a result of the manipulation, the designer can experiment with various design alternatives.

2. *Engineering Analysis.* Designers may also use the system's software programs to perform engineering analysis such as dynamic simulation and finite element analysis. Engineering analysis allows designers to determine the load and stress of complex structures quickly and accurately. Most CAD systems have the capabilities to perform mass analysis of properties on surface areas, weights, volumes, centers of gravity, and moments of inertia.

3. *Automated Drafting.* After the design has been developed, hard copy engineering drawings may be created directly from the CAD system's data base. Designers can perform such operations as automatic dimensioning, scaling, and developing sectional views. Engineering drawings can be made to adhere to company drafting standards by programming the standards into the CAD system.

A CAD system is a combination of hardware and software. The hardware includes a central processing unit (CPU), memory, and some types of input and output (I/O) devices. The CPU carries out program instructions to perform operations on data. Input devices include such devices as keyboard digitizer tablet and mouse, whereas output devices are usually plotters and graphic displays on CRT printers.

Computers are generally grouped into three main classes: microcomputers, minicomputers, and maxis (mainframe) computers. The computers in each of these classes of computers are subdivided into further classes. As a result, a supercomputer is found in each main classification; that is, there are super microcomputers, minicomputers, and mainframe computers.

Computers are classified according to their processing speed, accessible memory, and word length. A word length is the number of bits a computer can handle at a time. An internal word length is the number of bits the CPU can handle. An external word length is the number of bits that can pass a data bus at the same time. An internal word length determines the CPU's processing speed, and an external word length determines the amount of main memory the CPU can access directly. Internal and external word lengths are often the same, but they may differ depending on the system.

Microcomputers

A microcomputer used in CAD systems is usually a general purpose, single-chip CPU that has either 8-, 16-, or 32-bit word lengths. A byte is a grouping of adjacent binary digits operated on by the computer as a unit. The most common size byte contains 16 binary digits. Main memory sizes range from 64,000 to 2 million bytes. The newer generation of micros, often called supermicros, work with 32-bit word lengths and can access 10 million bytes of main memory.

Software to perform CAD/CAM applications is usually purchased separately from the hardware in order to convert the personal computer (PC) and compatibles to perform various manufacturing operations. The PCs have captured a large share of CAD/CAM applications. Such PCs usually have 16-bit or 32-bit CPU and can be purchased for $3,000 to $8,000.

Standard configurations of PCs are often not adequate for productive CAD/CAM operations. A substantial amount of internal read/write memory, often called random access memory (RAM), is required to run graphics-intensive tasks such as drafting, computation, simulation, and analysis. Processing speed can also be increased by adding a mathematical processor to the basic system.

External storage devices, such as floppy disk drives, are often needed in CAD/CAM applications. Many microcomputers require that CAD software be stored on two and in some cases as many as 10 floppy disks. Hard disk drives, which store 20 to 80 million bytes of data on rigid disks, are often used in micro CAD/CAM systems because of their capacity and access speed exceed those of floppy disk drives. A microbased CAD/CAM system is shown in Figure 3.1.

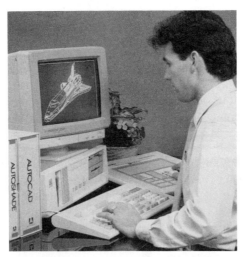

Figure 3.1 Microcomputer-Based CAD System

Minicomputers

Most minicomputers have 32-bit word CPUs, whereas mainframes are typically located in a corporate data processing facility or in the plant and serve as the host computer. Minicomputers are often placed near the engineering department as the host computer for the CAD/CAM system so that CAD/CAM does not rely on outside computing resources such as the corporate or plant computer. Most minicomputers support many different users and can perform many tasks at the same time. Minicomputers generally have the best price to performance ratio of any computer class. Many minicomputer vendors offer a range of processors, ranging from desktop models to superminis that perform as mainframes in speed and memory. Superminis, with operating speeds in excess of 3 to 10 million instructions per second (MIPS), are becoming increasingly important in CAD/CAM applications. As a result, these processors extend minis into the mainframe range but do not require special computer facilities, and the implementation, installation, operation, and maintenance costs are much lower than for a mainframe.

Another characteristic common to most minicomputers is a virtual memory operating system. Operating systems perform basic housekeeping tasks and permit other programs to run. Virtual memory operating systems automatically swap portions of a program between memory and disk storage. As a result, the computer can run large programs such as solid model or engineering analysis. A typical CAD system based on a minicomputer is illustrated in Figure 3.2.

Mainframes

Mainframe computers, often called maxis, are used in applications requiring substantial data processing and large memory capacity, such as company-wide data base management and financial operations. Mainframes are often used to provide

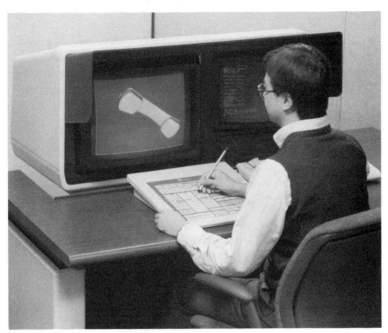

Figure 3.2 Minicomputer-Based CAD System

processing power after a problem has been set up with the aid of a micro or mini. The problem is then run in a batch mode from beginning to end with no interactive interruptions. Mainframe computers are also used to link smaller processors that are distributed throughout a company.

Supercomputers

A supercomputer uses the most advanced technology in processing techniques and memory organization to reach computing speeds many times that of computers in its class. Using a mainframe, CPU-intensive applications in CAD/CAM, such as simulations, finite element analysis, solid modeling, kinematics, and computation speed can be increased 100 times or more. Such speed makes supermainframes cost effective. Many complex problems in CAD/CAM that were not even considered several years ago can now be solved economically.

Supercomputers work faster than conventional computers because of data pipelining circuits and large internal memories. Conventional computers perform one instruction at a time on each individual element of data. In contrast, vector-processing hardware performs computations more rapidly by handling data in blocks or sectors. Data elements are streamed sequentially from memory to the vector processor, which is segmented to do multiple operations at one time. As a result, only one instruction is used to execute the same operation on all elements of a vector. This continuous flow of data is called pipelining.

Time-Sharing

An alternative to owning a computer is time-sharing, which provides an inexpensive method of accessing a mainframe or supercomputer. The user's terminal is connected by telephone lines to the computers at a service bureau. A customer pays only for the amount of computer time actually used.

Many CAD service bureaus provide consulting services and software customization for users. The cost of a time-sharing system includes training, equipment, a software usage fee, and consultation. The greatest cost of time-sharing, however, remains computer time. A firm may initially use CAD/CAM by time-sharing CAD services from a service bureau. The basic computer-aided drafting required by some users can be performed for a few hundred dollars a month. As the system becomes more heavily used, a firm then may decide whether to get an in-house system.

Workstations

Engineering workstations are personal computers (PCs) with improved computer processing power. Low-end workstations typically use 16-bit microprocessors. A hard disk is used in the workstation for permanent storage of data, and software is used for applications such as spreadsheets or word processing. High-end worksta-

Figure 3.3 Engineering Workstations

tions contain more powerful processors than personal computers. Typically, a 32-bit microprocessor with 0.2 to 10 MIPS, 1 million bytes or more of main memory, a hard disk of 30 million bytes or more, a high-resolution graphic display, and networking capability are available at high-end workstations.

As shown in Figure 3.3, a workstation can work as a standalone unit serving as a design and analysis station. More CPU-intensive tasks such as solid modeling can be uploaded to the larger host computers for processing.

Graphics Hardware in Computer-Aided Design

A modern CAD system is based on interactive graphics. Typically, a standalone CAD system includes the following hardware components:

- One or more design workstations that consist of a graphics terminal and an input device.
- One or more output devices
- A central processing unit
- Secondary storage

These hardware components could be arranged in a configuration such as that illustrated in Figure 3.4.

The Graphics Terminal

Users of interactive graphics systems communicate with the computer through graphics terminals. The computer translates an image drawn on the screen into a mathematical model and stores it in memory. The user may instruct the computer to retrieve data or drive a plotter for hard copy output of the drawing, or the user may instruct the computer to analyze the model. Results of such an analysis are also displayed on the screen.

Nearly all computer graphics terminals available today use a CRT as the display device. The operation of the CRT is illustrated in Figure 3.5. A heated cathode emits a high-speed electron beam onto a phosphor-coated glass screen. The electrons energize the phosphor coating, causing it to glow at the points where the beam makes contact by focusing the electron beam. Changing the beam intensity and controlling its point of contact against the phosphor coating through the use of a deflector system generates the image.

Most CAD/CAM systems use one of three main types of graphics terminals: raster scan, vector refresh, or storage tube. There are differences in characteristics such as cost, resolution, color, animation, and brightness. There are also plasma panel displays and liquid-crystal displays.

Raster scan terminals. Raster scan terminals operate by causing an electron beam to trace a zigzag pattern across the viewing screen. The operation is similar to that of a commercial television set. A television set uses analog signals to construct the image on the CRT screen; the raster scan terminal uses digital signals generated by a computer to construct the image.

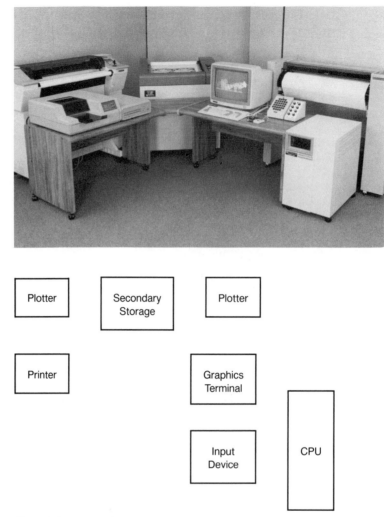

Figure 3.4 Hardware in a Computer-Aided Design System

Raster scan terminals produce images with a matrix of picture elements called *pixels.* A pixel is the short term for picture element. It is a small rectangular division of the video screen. Typically, each pixel is illuminated to one of 64 intensity levels to produce a viewable image with various gray levels. An image is drawn when the electron writing beam is swept, or rastered, across the entire screen line by line from top to bottom. A modulating signal changes the beam intensity at the proper points on the screen to selectively illuminate each pixel according to a pattern stored in memory, which may be in a separate computer or at the terminal itself. Each displayable point requires at least one corresponding bit in the memory. Collectively, these stored points in memory are called a bit map. A picture tube with 256 lines of resolution and 256 addressable points per line to form the image would require 256

Figure 3.5 Cathode Ray Tube

× 256, or more than 65,000, bits of memory storage. Each bit of memory contains the on/off status of the corresponding pixel on the CRT screen. This memory is called the frame buffer, or refresh buffer.

Resolution is an important factor when selecting display for CAD/CAM systems. Viewable resolution and addressable resolution refer to different capabilities of the raster terminal.

- *Viewable resolution* indicates the level of sharpness displayed on the CRT, and it is limited by the physical characteristics of the tube.
- *Addressable resolution* is the amount of memory allocated for image description.

In a standard raster terminal, addressable resolution equals viewable resolution. However, in more powerful raster terminals, the number of pixels stored in memory exceeds the amount that can be displayed on the CRT. This type of terminal can perform zooming and panning functions at the terminal.

Screen resolution is defined by the number of horizontal pixels and the number of vertical pixels. In CAD/CAM applications, resolutions of 1000 × 1000 are often used, but most CAD stations based on low-cost PCs use a 6400 × 400 screen. Some display monitors use an interlocked method of scanning in which the electron beam strikes every line on the screen. Scanning that is not interlocked can show motion with less blurring and less flicker, but it is more costly.

Drawing speed is the time needed to retrieve and display information on the screen. The time needed to repaint graphics information on screen depends on both resolution and rate of data transmission. Advanced graphics terminals have functions built in the hardware to process graphics information locally and thus reduce reliance on repaint speed and on host computers.

Vector refresh terminals. Vector refresh terminals use the stroke-writing approach to generate the image on the CRT screen. The electron writing beam is deflected to trace image lines directly on a screen in a continuous sweep. The phosphor elements on the screen surface are capable of maintaining the brightness for only a short time. For the image to be continued, the picture tubes must be refreshed by causing the directed beam to retrace the image repeatedly.

The major benefit of vector refresh terminals is their ability to display animation on the screen. The image is very bright, and the resolution is excellent because images are produced by a continuous sweep of the writing beam instead of a dot matrix. It is also possible to erase partial content of the screen.

The major disadvantage of vector refresh terminals is that on diversely filled screens, the image appears to flicker. Because additional hardware and memory are required for image storage and refreshing, vector refresh terminals are generally more expensive than other types of graphics terminals.

Storage tube terminals. *Storage tube terminals* are also called direct view storage tubes, or DVSTs. The term *storage tube* refers to the ability of the screen to retain the image. A stroke-writing beam energizes the phosphor elements on the screen. The flood gun electrons that constantly spray the entire screen continue to illuminate the phospher along the track that has been energized by the stroke-writing beam.

The major advantage of a storage tube CRT is that large amounts of data, either graphical or textual, can be displayed on the screen without flickering mainly because the image is stored on the screen rather than being constantly refreshed from computer memory.

The major disadvantage of a storage tube CRT is that selective erasure is not possible. Once an image is displayed and stored, it cannot be changed without repainting the entire screen, making interactive manipulation and animation impossible. To overcome these limitations, some storage tube terminals have a feature called write-through to permit editing before the image is stored. These terminals use a low-energy writing beam and produce an image without storing a change on the screen. This unstored image is then retraced by the writing beam in a manner similar to that of a vector refresh terminal. The operator may selectively erase portions of the image and dynamically manipulate it. After the image is complete, a high-energy beam stores it on the screen.

Storage tubes have no color capability. Their low brightness and contrast require dimmed ambient lighting and screen shielding.

Plasma display panel. The plasma panel display is an interactive display that is not based on CRT technology. A plasma panel is an array of small neon bulbs, each of which can be switched on or off. A typical plasma panel has 64 cells or bulbs per inch and is 8 inches square with 250,000 cells.

The plasma panel provides a display of medium resolution that does not require a refresh buffer. The bulbs are fabricated as part of an integrated panel made of three layers of glass. The outside layer has thin vertical strips for electrical conductors. The center layer has holes that form the bulbs, and the inside surface of the other layer has thin horizontal strips for electrical conductors. A voltage difference on the line is used to fire the neon in the bulb and make it glow. A lower voltage is used to turn off a bulb. The bulbs can be switched on or off in about 20 microseconds.

The major benefit of the plasma display is that the panel is flat, transparent, and rugged. It needs no refresh buffer and can be used on portable computers or display terminals where size and weight are important factors in terminal design.

Liquid-Crystal Display System (LCDS). Some portable computers use an LCD display (Figure 3.6) because it is not a light source but reflects ambient light in the desired pattern. Its power requirements are very low. Color filters can be used in an LCDS to provide color. The Data General One Personal Computer uses an LCD display. A very popular personal computerbased CAD system, AUTOCAD, runs on the portable computer without modification. The lightweight portable can be used in areas such as field service, where personnel can take the machines to the customer's sites.

Figure 3.6 Liquid-Crystal Display

Graphics Input Devices

Several kinds of input devices are used to interact with the graphics workstation to facilitate communication between the user and the system.

Keyboards. Keyboards are of two basic types. The alphanumeric keyboard can be used to enter commands, text, and parameter values. The programmed function keyboard has a number of pushbuttons. Sometimes it is designed as a separate unit, but more often the buttons are integrated with the main keyboard. The function of each key is usually preprogrammed by the programmer. The operator can start an activity such as the notation of a displayed object by depressing a button and can then terminate the activity by releasing the button.

Light pens. Light pens were developed early in the history of interactive computer graphics. The light pen does not emit light to draw a picture. It senses or detects light from the picture element on the screen. A light pen consists of a pencil-sized cylinder with a light sensor in one end. The other end is connected to the computer by a cable. As the user positions the light-sensitive pen tip to select a point on the screen, the light pen sends a pulse when this screen is bright. The pulse produced by the light pen is then used to calculate the screen position where the pen was. Most light pens also produce a second signal, which the user generates by pushing a button on the pen to signal to the computer the user's selection of a point.

Track balls. Track ball controls are large plastic balls mounted with some fraction of the ball surface protruding from the top of the enclosing unit. Rotating the ball moves the screen cursor. Track balls are often used to select commands displayed on the display screen. The ball positions the cursor in the area of a menu. Pushing a keyboard key makes the selection. The computer reads the cursor position at the time the key is pressed and executes the procedure listed in the menu.

Touch screens. Two kinds of touch screens—mechanical and optical—are used in CAD/CAM systems. The mechanical touch screen is a transparent screen overlay that detects the location of the touch. Optical touch screens use rows of light emitters and receptors mounted just in front of the screen, with the touched location determined by which beams are broken.

Joysticks. Joysticks, like thumb wheels, mice, and track balls are potentiometric devices. The user controls the motion of a cursor on the screen by pushing the joystick in the direction of desired motion. Many users prefer joysticks because they allow rapid cursor movement for relatively small displacements of the device, enabling graphic operations to be performed quickly. Some joysticks have a third degree of freedom: Pushing the handle up and down or twisting it provides data entry in the z axis.

Mouse. A mouse is a small hand-held device with an attached wire (Figure 3.7). The mouse can be moved around by an operator on any flat surface to provide

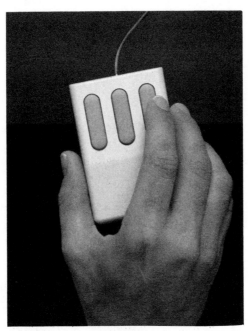

Figure 3.7 A Mouse Input Device

graphic input. A mechanical mouse must be operated on any flat surface, whereas an optical mouse needs a special reflector plate on which to run. The accuracy of the mouse is not that of a tablet. The advantages are that a mouse requires only a small desk space and is generally less costly to purchase, install, and operate than a tablet.

Digitizers. Digitizers (Figure 3.8) consist of two basic elements: (1) a locator in the form of a pen, puck, or cursor; and (2) a tablet, the size of which can range from just a few square inches to the size of a drafting table.

Most tablets use some electrical sensing scheme to measure the position of the stylus on the cursor. In one design, a grid of wires is embedded in the tablet surface. The electron magnetic coupling between the electrical signals in the grid and a wire in the stylus, or cursor, induces an electrical signal in the stylus. The strength of the electromagnetic coupling can be used to indicate the physical motion of the puck or cursor. The electrical signals are then routed to the computer and displayed on a CRT screen.

Data can be picked up by the cursor in two general modes. In the point mode, the user locates specific points with the cursor and then presses the appropriate function buttons to enter the data. A typical operation would be to enter the end points of a straight line. In the stream mode, a continuous flow of data points is sent to a processor, with the data rate determining the digitizer resolution. This method is generally used for entering curves and irregular lines.

In applications for which three-dimensional data are necessary, a 3D digitizer (x, y, and z coordinates) can be used. The user moves an arm, or wand, along the

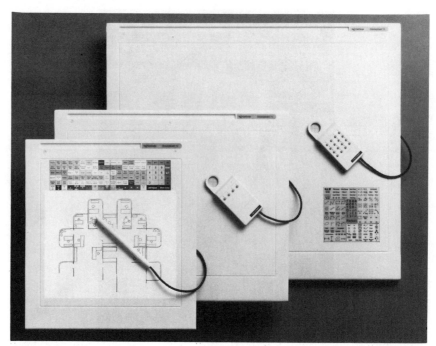

Figure 3.8 A Typical Digitizer Used in CAD Workstations

surface of a three-dimensional model. The tip of the arm picks up the three-dimensional information and feeds the coordinates into the computer for further analysis.

Some digitizers use sound pulse to couple a stylus with microphones along the sides of the tablet. A spark is generated at the tip of the stylus, and the time from the occurrence of the spark to the arrival of its sound at each microphone is used to calculate the distance of the stylus from each microphone. An advantage of some digitizers is that they receive no wires and therefore no tablet, making them portable. The *x* and *y* strip microphones can be placed on any surface.

Speech recognizers.   Speech recognizers can accept verbal commands and values as input. Word recognizers must be trained to the characteristics of an individual's speech. A word is filtered into several frequency bands, and patterns based on the relative loudness or pitch of the sound are matched against a dictionary of recorded patterns.

Scanners.   Scanners are used for entering entire existing drawings into the CAD database. A typical system contains a scanner and a CAD workstation for viewing and editing the drawing. The scanning system scans a document to create a raster image. The scanning is usually measured in dots. Some CAD applications require a more editable file, in which case raster data are often converted to vectors. Sophisticated software is required to overcome the many difficulties in this conversion process. In the traditional scanning method, in which the scanner scans the entire full-size drawing, scanning a large E-size (48″ × 36″) drawing may take up to an hour. More innovative methods convert the drawing into microfilm. The scanner then scans the image on the microfilm. This process greatly reduces the time needed to scan the entire drawing.

Graphic Output Devices

A CAD/CAM system is not complete unless it can make hard copies of design or analysis created on the screen. Hard copy devices include electrostatic plotters, serial thermal transfer printers, page thermal transfer printers, ink jet printers, impact dot matrix printers, laser printers, pen plotters, and CRT-based film recording systems.

Electrostatic plotters. Electrostatic plotters selectively charge a specially coated paper that attracts the toner to charged areas and creates an image. Electrostatic devices work at speeds up to 1 inch per second and can handle the larger format demands of CAD (Figure 3.9). A high-speed electrostatic device is able to complete an E-size plot in less than a minute.

Serial thermal transfer printers. Thermal printers use a heating mechanism to move a head that transfers dots of ink from a ribbon to a special nonband, smooth paper. The advantages of serial thermal transfer printers are a relatively low cost,

Figure 3.9 An Electrostatic Plotter

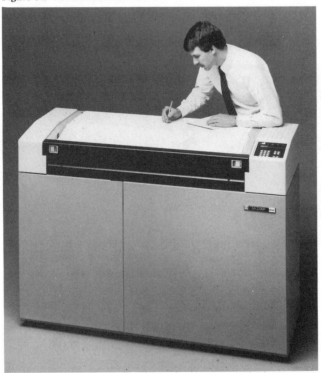

good color quality, and low noise levels. The disadvantage is their slowless. They often take up to 15 minutes to print one full-color page.

Page thermal transfer printers. Page thermal transfer printers operate like serial thermal transfer printers, except they print a full line of dots simultaneously (Figure 3.10). Monochrome units can produce hard copy up to six pages per minute. Because of anticipated developments, such as lower prices, improved resolution, greater speeds, and page thermals, transfer printers could become very popular for computer graphics users.

Ink jet printers. Ink jet printers create images and characters by projecting fine drops of ink onto the print medium (Figure 3.11). The drops can be continuous or pulsed. In a continuous system, the ink drops are selectively charged. Uncharged drops hit the print media and charged droplets return to a reservoir. In a pulsed printing system, the drops are selectively projected in a matrix pattern onto the print medium. Continuous jet operation makes a higher resolution image but requires an ink recirculation system. Ink jet printers are relatively slow, and they often develop ink leaks because of their complicated plumbing systems. Ink clogging is another problem.

Impact dot matrix printers. Dot matrix printers use a group of small pins to strike a ribbon (Figure 3.12). The resulting pattern comes out in the form of a character or a number. Multiple ribbons can be used for color printing. Impact dot matrix printers have their drawbacks; they are slow and noisy. However, much has

Figure 3.10 Page Thermal Transfer Printers

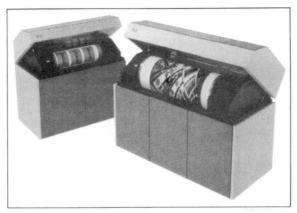

Figure 3.11 Typical Ink Jet Printers

been done to quiet these devices both through isolation boxes and redesigning the printing operation.

Laser printers. A laser printer uses a computer-controlled laser beam to create images on a photoreceptor. Images are than xerographically transferred directly to paper.

Pen plotters. Two basic types of electromechanical pen plotters are used in CAD/CAM systems. In a flat bed plotter, servocontrolled pens are moved in two axes over flat stationary sheets of paper. In a drum or roll plotter, pens, styli, or ink jets remain stationary or move along one axis while the paper moves in another axis on a revolving drum. A pen plotter, such as is shown in Figure 3.13, offers outstanding line and color quality for line drawings and is inexpensive to buy and operate.

Film recording systems based on CRT. Film recording systems offer the highest resolution and the truest color reproduction of all hard copy devices. In some applications, a camera is mounted directly in front of the CRT to take a picture of the screen image. In more advanced systems, special devices are used to photograph images directly from signals from the CRT. A film recorder is shown in Figure 3.14.

Figure 3.12 An Impact Dot Matrix Printer

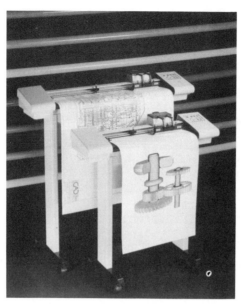

Figure 3.13 A Pen Plotter

Analysis and Optimization

Many CAD systems have analytical capabilities that allow parts to be evaluated with data to create a system model. In the traditional design process, component designs are based on load estimates modified by prototype testing. Using the simulation capabilities of the CAD system, designers can evaluate the overall product considerations, such as weight distribution, vibration, stability, and stress concentrations, without building a physical model.

Figure 3.14 A Film Recorder

Finite Element Analysis

Finite element analysis (FEA) is a computer-based technique for determining stresses and deflections in a structure too complex for classical analysis. The method divides a structure into small elements with easily defined stress and deflection characteristics. Many commercial programs for FEA are available. These programs require that the user know only how to prepare program input properly. An FEA is illustrated in Figure 3.15. The finite element method is applicable in several types of analyses such as static analysis, natural frequency analysis, transient dynamic analysis, heat transfer (thermal) analysis, and motion analysis.

Static analysis. In static analysis, designers determine deflections, strains, and stresses in a structure under a constant set of applied loads.

Natural frequency analysis. Natural frequency analysis calculates the stress vibration of natural frequencies and associated mode shapes of a structure. Designers use the result to predict the critical operating conditions for machinery.

Transient dynamic analysis. Transient dynamic analysis determines the history of a structure's time responses. Once this history is known, complete deflection and stress information can be obtained.

Heat transfer analysis. Heat transfer (thermal) analysis involves the temperature distribution in a structure where the thermal loads and boundary conditions are known. It can be used to solve problems of steady state and transient heat transfer. A thermal analysis is shown in Figure 3.16.

Figure 3.15 Finite Element Analysis Model

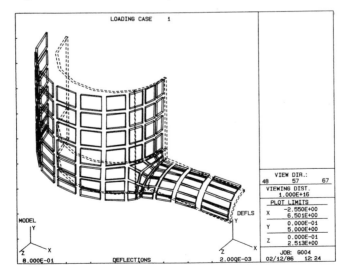

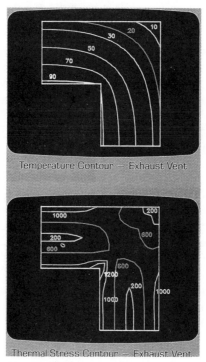

Figure 3.16 Thermal Analysis

Motion analysis. Kinematics is the calculation of geometric properties such as displacements, velocities, and acceleration for the synthesis of mechanisms required to produce a given motion. Traditional methods of kinematics are tedious, time-consuming, and costly. A CAD system with kinematic synthesis capability dramatically can reduce the time required for linkage design. A designer can visualize many design alternatives and refine the design.

Models

Geometric Modeling

The most important feature of a CAD system is the geometric model, representing part size and shape in the computer. The geometric model can be used to create a finite element model of the structure for stress analysis, or it can serve as input for automated drafting. It can also be used to create numerical control (NC) tapes for marking parts on automated machine tools or for producing process plans outlining the steps required to make the part.

Wire Frame Models

Most CAD systems are a form of modeling called wire frame modeling. A wire frame model is shown in Figure 3.17. The image assumes the appearance of a frame

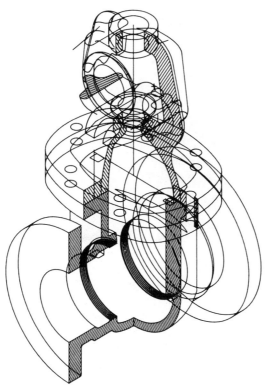

Figure 3.17 A Generated Wire Frame Model

constructed out of wire—hence, the name. The wire frame models can be ambiguous in representing complex physical objects. This limitation may be improved by blankout of dark hidden lines.

Solid Models

When the definition of structure boundaries is critical, many ambiguities of wire frame models can be overcome with surface models that define the outside part geometries precisely and help produce NC machining instructions. A major drawback of surface modeling is that it provides no inherent information about mass properties. Thus, a surface model cannot serve as a basis for engineering analysis programs, such as finite element analysis, that require information about part material, moments of inertia, centroid, and other properties that affect how the part responds to loads.

Solid modeling. Solid modeling overcomes the drawback of both the wire frame and surface models by defining parts as solid objects. A solid model is shown in Figure 3.18. When this model is used, computations of parameters such as weights and moments are possible, and cross sections can be cut through the model to expose internal details with minimal user interaction.

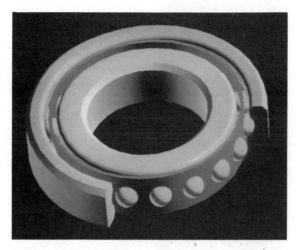

Figure 3.18 Solid Model

Solid models are constructed in two basic ways.

1. Constructive representation, also called C-rep or the building block approach. In this approach, the geometric primitives are combined using Boolean logic operations (union, difference, or intersection) to create a new shape.

2. Boundary representation (B-rep). Boundary programs build models by piecing together surfaces that enclose the spatial surface boundary of the object. The user constructs the two-dimensional view of the object, then uses sweep operations to move the two-dimensional surface through space to trace out a solid volume. A linear sweep translates the surface in a straight line to produce an extruded volume. A rotational sweep produces a part with axial symmetry. A compound sweep moves a surface through a specified curve to generate a more complex solid. Gluing joins two previously created solids with a common surface.

The two approaches have their relative advantages and disadvantages. Most industrial parts can be modeled by using the constructive solid geometry method, but components with complex contours, such as turbine blades, are more easily modeled by using the boundary method.

Automated Drafting and Documentation

For most companies, automated drafting is the first step toward total automation. Automated drafting equipment typically includes a microcomputer or a minicomputer, a monitor, and a plotter. A typical drafting arrangement is shown in Figure 3.19. Automated drafting systems require little or no knowledge of computers or programming. The user simply communicates with the system interactively with pictures instead of numerical data. Generally, a two- to six-fold increase in productivity is achieved by using automated drafting installations. The more com-

Fundamentals of Product Processes and Operations

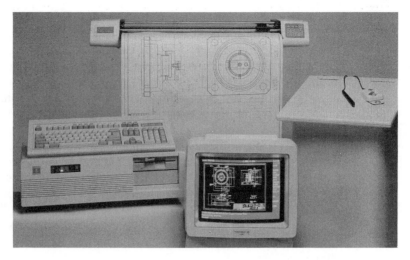

Figure 3.19 Automated Drafting

plex the drawing, the more advantageous is automated drafting. The more sophisticated drafting software becomes, the easier it is to use. And advances in input devices are making automated drafting smoother and faster.

Manufacturing Engineering

As discussed in Chapter 1, engineering systems are used extensively in manufacturing engineering. Once the design is complete, the feasibility of producing it economically is verified by a manufacturing engineer. Manufacturing engineering plans where, how, and when to perform work necessary to produce a product. It also coordinates internal and external orders, delivery dates, workmen, machines, and the like, thereby promoting efficient operation. This function may be divided into four responsibilities:

1. It provides producibility information to product design.
2. It develops process plans to make the product.
3. It determines specification and design of tools, jigs, and fixtures used to make, test, and inspect the product.
4. It solves production problems that develop during production or seeks out the people who can solve them.

Process Planning

Process planning determines the sequence of individual manufacturing operations needed to produce a given part or product. (Process planning is discussed further in Chapter 5.) The result of process planning is a detailed list of steps describing each operation and associated machine tools. This task is traditionally performed by a manufacturing engineer, an industrial engineer, or a process planner. The process planning procedure is very much dependent on the experience and the judgment of the planner. Because individual planners have their own opinions about what constitutes the best coating, there are differences among the operation sequences developed by various planners. Computer-aided process planning (CAPP) attempts to fully automate the process planning function on the basis of only CAD and the machining database.

Process planning and other manufacturing information are produced automatically by the computer from the CAD and machining databases for use on the factory floor. Much of the information needed is in these databases or a related data base at this time. If additional data are needed, they are put into the database by means of the computer. A typical example is by "filling out" a tabular chart that is displayed on a terminal. The data generated by similar systems and under similar circumstances can then be retrieved from the database to reduce unnecessary efforts.

Computer technology speeds up and simplifies process planning. It aids the process planner in managing and retrieving a great deal of data and a host of documents encompassing machinability data, machine specifications, tooling inventories, stock, standards, and already existing process plans.

Numerical Control

Numerical control may be defined as a form of programmable automation in which a process is controlled by letters, numbers and symbols. That is, NC is a technique of controlling machine tools with prerecorded, coded information (see also Chapter 4). An NC system consists of three major components (Figure 3.20) as follows:

1. The program of instruction
2. The machine controller
3. The machine tool

This program of instruction is discussed in this section.

Manual Data Input

At the machine tool level, the NC program can be input manually to the controller unit. This method is called manual data input (MDI) and is appropriate for only relatively simple jobs.

Numerical control programs are prepared by a part programmer who prepares the planned sequence of events for the operation of a numerically controlled machine tool. An NC program is a specific set of data and instructions written in source languages for computer processing or written in machine language for manual programming for the purpose of manufacturing a part of an NC machine. A typ-

Figure 3.20 Computer Numerical Controlled Machine Tool (CNC)

ical NC part programming manuscript is shown in Figure 3.21. In manual programming, the part programmer specifies the machining instructions on the programming manuscript. The form of these manuscripts varies because of such factors as the number of axes of the machine tool, format to be used, job to be done, programming technique and programming language used.

Computer-Assisted Part Programming

In the more complicated applications, manual part programming is an extremely tedious task that is subject to errors. In these instances, a computer is usually employed to assist in the process. Programming languages such as Automatically Programmed Tools (APT) or COMPACT II enable the user to develop NC instructions more quickly with fewer errors.

Numerical Control Programming with Interactive Graphics

Interactive computer graphics techniques are being used at the machine tools, programming workstation, or with CAD/CAM systems to simplify the programming process.

The process begins by defining the part geometry. This computer representation of the part geometry (geometric model) may be transferred from the CAD database, or it can be built by the user from scratch at the NC workstation. After the part program has been created from this geometric model, tool paths can be displayed over the part to check for errors. More powerful systems can be designed to present an animation of tool motion as further check on tool paths. Such simulations virtually eliminate machining problems on the shop floor.

| NC machine programmed information for use with NC rotating work | | | | | | | | Date | Prog. no. | Tape no. | Lathe |

OPERATION DESCRIPTION	Sequence no. N	G func. G	Dimension X	Dimension Z	I	K	Feed-rate no. F	Speed no. S	Tool no. T	Misc. Func. M
Move tool to initial position.	N001	G01								M12
Select tool position in boring turret to ensure safe operation — dwell 2 seconds.	N002	G04	(X02)						T10	M24
Select turning turret — ensure tool is in position.	N003									M23
Dwell to select tool, speed and start spindle clockwise.	N004		(X02)					S07	T11	M03
Position turning tool at rapid traverse by moving 45mm in X towards the workpiece.	N005	G01	X − 045				F			
Position at rapid traverse in Z-45mm towards workpiece.	N006			Z − 045			F			
Move in to depth of cut in X at fine feed for 5mm.	N007		X − 005				F			
Feed to workpiece in Z 5mm and continue to 70.5mm length of cut.	N008			Z − 0755			F			
Rapid out to initial position in X.	N009		X + 050				F			
Rapid back to initial position in Z. Program stops to enable operator to measure workpiece.	N010			Z1205			F			M00
Restart program.	N011									M17
Select position 1 on turning turret to ensure safe operation.	N012	G04	(X02)						T10	
Select boring turret.	N013									M24
Dwell for tool change. Change tool — select speed — change speed.	N014		(X02)					S14	T62	M03

Figure 3.21 A Part Programming Manuscript Form

Although NC programming is still the workhorse in CAM, computer software programs are being developed for specialized tasks that can be defined as manufacturing design and analysis. This software covers the areas that are neither product design nor pure production. It assists in intermediate tasks required to manufacture a part. Jobs that fall into this category include the creation and analysis of models, dies, tools, and fixtures. The software for designing these items is different from that used to assist in creating the manufacturing process—for example, process planning, offline robot programming, and factory simulation.

Mold Design

The development of molds traditionally relies on analysis programs not integrated into a CAD/CAM database. For example, injection molding projects typically begin when a product designer generates a series of sketches for the approval of the concept by marketing. A conventional orthogonal drawing is then produced from these sketches. Projected cost, manufacturing processes, tooling, and marketing strategies are developed. The drawings are also used to reduce each of the steps in the iterative development cycles.

To reduce development time and cost, CAD/CAM software programs have been developed to simulate flow into the mold, cooling, runner design, and NC machining so that costly tooling can be made right the first time. Manufacturing software is either developed by the CAD/CAM vendor or the manufacturing enterprise. It may also be bought from a third-party vendor and integrated into the CAD/CAM system. The software then becomes another module in the CAD/CAM system, using the same database for tasks such as geometric modeling and NC programming.

Die Design

Die design includes forging dies, progressive dies, and many others. Progressive dies entail strips of metal being run through a series of stations, with a feature added at each station. The last station cuts off the finished part. Using a CAD/CAM system, the designer performs his tasks by using a set of rules and procedures written in a familiar system lingo such as "defined punches." Many of the required drawings are produced automatically once the design is complete. Output is created for running the NC machines to produce the mold, die, as well as a bill of materials.

Designers of forging dies can enter forging data such as material specifications and the geometry of parts. The designers can also enter hammer size and type into the program and can alter the output to suit their forging process in an attempt to optimize efficiency. The software then calculates estimations of stock size and material yield, "flash land," geometry, flash force, and total energy requirements. The program can also calculate the number of flash-forming die impressions required and the number of forging blows needed.

Fixture designers can simplify their jobs by using on-line catalogs of fixtures from suppliers. By using fixtures instead of a part, designers can perform NC programming more realistically without having to modify the program on the shop floor.

Quality Control Planning

The increased automation of design and production methods and processes demands that designers consider the product's manufacturability so that the product line will meet specifications and perform according to requirements. The design and manufacturing functions' willingness to exchange sufficient design information and to compromise can result in improved product quality, reduction of parts inventory, fewer engineering changes, and a shorter, less costly production cycle.

Quality Designed into Products

Today's product designer must have knowledge of materials and manufacturing processes and their costs. To keep costs down, designers need to learn as much as possible about new materials and their application in product and process design. A knowledge of the manufacturability limits (limits in precision, surface finish, and the range of materials that can be machine molded and processed) of manufacturing equipment is also critical.

Products designed for manufacturability and automation result in a reduction of parts and inventory and reduction in engineering changes, assembly time, and costs. The consequences are lower costs and better quality for the product.

Quality Engineering

Quality engineering (QE) is an engineering discipline for the establishment of quality tests, inspections and quality acceptance criteria, and the interpretation of quality data. This discipline is also a support function to production engineering test and to inspection equipment design and selection. Quality engineers support product design engineering in determining methods and procedures for testing and inspecting products to make sure that the products will meet all the designed requirements.

After test points have been established, the design of the test equipment proceeds. While the test equipment is being produced, test procedures and instructions are being prepared. Final quality assurance (QA) accepts (buy-off) the test equipment and procedures is coincident with the buy-off of the first production item tested. Quality assurance is a broad term used to include both quality control (QC) and QE.

Engineering Support

Engineering support is provided to many functions throughout the complete manufacturing cycle. Activities on a specific product in many manufacturing enterprises begin long before a manufacturing authorization (MA) is issued. An MA is a document, having the same meaning as work authorization, used to authorize performance of work within the production function.

Computer-Aided Engineering in Manufacturing

Before computers and interactive graphics systems became a commonplace in manufacturing, engineers designed products by manually calculating loads and stresses in the most critical areas of key parts. The results of their designs relied heavily on approximations. Designers tend to beef up the structure to be extra safe. Such a product development process typically yielded overdesigned structures, and designers tended to avoid exotic designs and stick to familiar standard configurations.

The use of computers in engineering has radically changed the design process as well as the resulting designs. Using techniques such as finite element analysis and dynamic simulation, the designer can design the strength of the structure to precisely match the expected operating loads. The engineers can now develop more complex and innovative designs that were formerly impractical.

Designers may rely too heavily on computer analysis and attempt to develop designs that barely exceed maximum expected loads. Slight variations in the manufacturing process or in materials properties or loads far greater than those normally encountered in ordinary service can result in product failure. Thus, designers must be extremely familiar with the design and the programs used in its development to spot errors in results.

Exercises

Self-Study Test

1. Define the term *super microcomputer.*

2. Define the term *engineering workstation.*

3. What was the most important event in the coming of age of the supermicro-computer in CAD/CAM applications?

4. What are the main differences between a minicomputer and a mainframe computer?

5. What is computer-aided design and drafting?

6. List the most frequently used input/output devices.

7. What is the difference between a master graphics display and the vector refresh graphics display?

8. Define CAD, CAM, CAE, and CIM.

9. Explain the benefits of using CAD in mold and tool design.

10. What are the advantages of NC? What are the disadvantages?

11. What are two responsibilities of the CNC part programmer?

12. Explain the concepts of design for manufacturing.

References

[1] Burstein, J., *Computers and Information Systems.* Holt, Rinehart and Winston, New York, 1986.

[2] Chang, T. C., *An Introduction to Automated Process Planning.* Prentice-Hall, Englewood Cliffs, NJ, 1987.

[3] Goetsch, D. L., *Fundamentals of CIM Technology.* Delmar, Albany, 1988.

[4] Grover, M., *Automation, Production Systems and Computer Integrated Manufacturing.* Prentice-Hall, Englewood Cliffs, NJ, 1987.

[5] Kaal, Irvin, *Numerical Control Programming in APT.* Prentice-Hall, Englewood Cliffs, NJ, 1986.

[6] Ludema, K., Caddell, R., and Atkins, A., *Manufacturing Engineering—Economics and Processes.* Prentice-Hall, Englewood Cliffs, NJ, 1987.

[7] Pao, Y. C., *Elements of Computer-Aided Design and Manufacturing.* Wiley, New York, 1984.

[8] Plossl, K., *Engineering for the Control of Manufacturing.* Prentice Hall, Englewood Cliffs, NJ, 1987.

[9] Preston, E., Crawford, G., and Coticohia, M., *CAD/CAM Systems.* Dekker, New York, 1984.

[10] Seames, W. S., *Computer Numerical Control.* Delmar, Albany, NY, 1986.

[11] Stover, R., *An Analysis of CAD/CAM Applications.* McGraw-Hill, New York, 1985.

[12] Vail, P. S., *Computer Integrated Manufacturing.* PWS–Kent, Boston, 1988.

4

Computer-Aided Production Systems

Overview

The production function, often equated with manufacturing, embraces the implementation of the product design, the process, the machines, the material handling, and the energy needed to actually produce a product. The product is fabricated, assembled, and tested to make sure that it meets all the design specifications. That is, the production function's primary responsibility is to change the shape, composition, or combination of materials, parts, or subassemblies to increase their value so that the finished product has utility in the common marketplace. Production is concerned with the factory floor and its related activities.

Such responsibilities require production to use many production and control subsystems (PACS) as discussed in Chapter 2. The PACS, often called basic functions, are selected from the major manufacturing functions to form a production entity that is designed to carry out the interrelated production activities and operations with a high degree of efficiency and effectiveness. Typical production and control subsystems are scheduling, parts assembly, testing and inspection, materials handling, inventory control, production management—just to name a few.

Productivity in major manufacturing industries declined dramatically in the 1980s. Some contributing factors to the decline in productivity are government regulation in price controls; the Environmental Protection Agency (EPA) and the Occupational and Safety Health Administration (OSHA); powerful labor forces; high cost of labor and materials; high indirect costs; management's apprehensive acceptance of new production technologies; human limitations; and logistics management. Reckoning with such factors requires the application of productivity-enhancing factory automation measures.

Computers in Computer-Aided Manufacturing (CAM)

Computers' ability to receive and handle large amounts of information, coupled with their fast processing time, make a systems approach to production indispensable. As a result, computers are taking increasingly important roles in computer-aided production and control systems (CAPACS). The application of computer technology to manufacturing production has become a reality in today's manufacturing environment. Computer-aided production and control systems include interactive graphics system (IGS), computer-aided process planning (CAPP), computer-aided testing (CAT), and computer-aided inspection (CAIN), numerical control (NC), flexible manufacturing systems, (FMS), and robotics. These systems are generally under the control of minicomputers or supermicrocomputers.

The application of CAPACS to the actual conversion of raw materials and shapes to the end product is referred to as Computer-Aided Manufacturing (CAM). A conceptual CAM hardware architecture is illustrated in Figure 4.1. This architecture shows a supermicrocomputer being used as an area controller. At this level, the area controller supervises and coordinates a group of stations to perform related manufacturing production activities. The controller also transmits data collected at the station (operation) level to the CAM host computer.

The CAM host computer is a satellite computer in the manufacturing system and serves in a managerial capacity for production. The host computer manages the activities of area controllers under its commands. Data are collected, processed, analyzed, and stored for other applications from individual stations, as shown in Figure 4.2. The host also sends instructions down to the separate processes and stations. The station supervisor coordinates the processing activities of each station under its command.

Figure 4.1 A Conceptual CAM Hardware Architecture

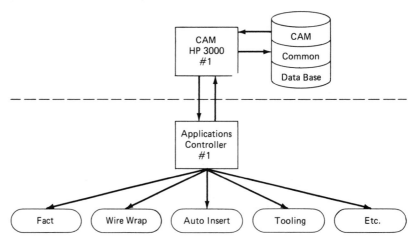

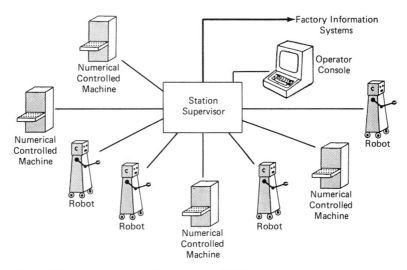

Figure 4.2 Station Supervisor Computers in CAM

CAM Functions

Computer-aided manufacturing embraces and provides assistance to many production functions, a number of which are illustrated in Figure 4.3. A CAM system design should be such that interaction between all functions is effective and efficient. All CAM functions should be integrated with the host computer (see Figure 4.1) by a common file system, usually referred to as a common database system.

With a well-organized database system, all information relating to an issue can be obtained, not just from a few functions, but from all functions in a matter of moments, thus providing management with a powerful tool.

Figure 4.3 Typical Functions in Computer-Aided Manufacturing

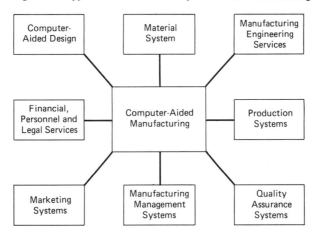

In true CAM systems, computers control the entire production operation. Workpieces are palletized and stored under computer direction. Management generates a call to the computer for a workpiece to be made. The automatic storage/retrieval system (AS/RS) delivers the material to a conveyor system, on to which a robot loads it. At a proper location, another robot takes it off and puts it on another conveyor that will take it to the workcell where it will be machined. Other operations necessary to add additional value to the workpart will be done by other automated systems.

Fabrication and Assembly Processes

Fabrication and assembly processes include many methods of modifying the shapes and properties of materials or of assembling two or more fabricated parts so that they fit each other in a predetermined relationship.

Starting at the bottom level of a CAM data processing facility on the factory floor are the dedicated processing equipment and machine controller (Figure 4.2). This machine controller is under the control of an area controller (the supervisory computer), as shown in Figure 4.2. The supervisory computer supervises a cell, which consists of a group of workstations. The cell controller controls the processes and communicates only with its supervisor at a higher level. The cell supervisor transmits information related to the process and operation to the host computer at the function level. Thus, the dedicated controller at the process level controls the tool, and such variables as feed speeds, cutting rates, processes, and related manufacturing operations. Production data from this level of processes and operations are transmitted to the cell supervisor for further processing.

Many activities take place on the factory floor. Typical types of machines and equipment used in carrying out these activities include numerical control (NC), robot, testing and inspection, data collection, material handling and transfer, and various types of assembly machines and equipment. Each cell is equipped, automated, and integrated to carry out various functions or processes, resulting in the creation of islands of cell automation. Most of these islands are composed of proprietary systems that cannot communicate or interact easily with systems from other vendors.

Thus, the key to computer-aided production (CAP) systems includes more than the machines, equipment, and processes on the factory floor. They encompass systems integration through a common data link, an intelligent data base system, support, training, consulting, customization, and a reliable data communication system.

Industrial Robots

Developments in robot technology have greatly affected the fabrication and assembly processes on the factory floor and offer potentials for improved productivity never before possible.

Robots may be designed to perform a variety of tasks in many application areas. The Robot Industries Association (RIA) defines the industrial robot as a reprogrammable multifunctional manipulator designed to move material, parts, tools, or specialized devices through variable programmed motions for the performance of a variety of tasks. The RIA uses this definition for a robot in an attempt to clarify which machines are simply automated machines and which machines are truly robots.

Properly programmed, an industrial robot can perform many manufacturing tasks. With preprogrammed routines, a robot can trace the three-dimensional image of an object, locate corners, find holes, separate multiple objects, and identify an object. These routines can also enable the robot to specify the grip points and the orientation of a workpiece. Robots have become very "intelligent" through the application of artificial intelligence (AI), which is discussed in Chapter 5.

Intelligent robots have recently moved into factories from research laboratories. These robots can "see" with vision cameras and "feel" with sensors and pressure transducers. A transducer is a device that converts energy from one form to another. These robots can move their fingers, arms, and legs using feedback control circuits. By using computer programs that operate on sensory inputs, they can "think" in a functional sense in order to make their own decisions without continuous human control.

Kinds of Industrial Robots

Current industrial robots may be viewed as consisting of three major functional components, illustrated in Figure 4.4: (1) manipulator, (2) power source, and (3) controller. A fourth component found in some robotic systems is the end effector (discussed on page 119). On the basis of the major components, one can classify industrial robots in a number of ways.

The manipulator. The manipulator is the arm of the robot. It is used to do the physical work by allowing the robot to bend, reach, and twist to pick up materials, parts, or special tools used in production. Thus, one way of classifying an industrial robot is by its manipulative functions. These include the pick-and-place models with a simple gripper design; continuous path robots that perform special jobs such as spraying and coating; and those that can be used to perform many types of manipulative functions.

The movements of the robots are often called the degrees of freedom of the robot. Industrial robots can have from 3 to 16 manipulator axes. Robots used in industrial application most frequently have a range of two to six axes. Figure 4.5 shows a robot with six axes. This Cincinnati Milacron 6CH Arm can lift a 175-pound load and move it at 50 inches per second.

Most industrial robots are generally capable of executing six basic types of movements, excluding the action of the end effector for example, the hand or gripping device illustrated in Figure 4.5. Three of these major axes are arm and body movements; and three wrist movements are called minor axes.

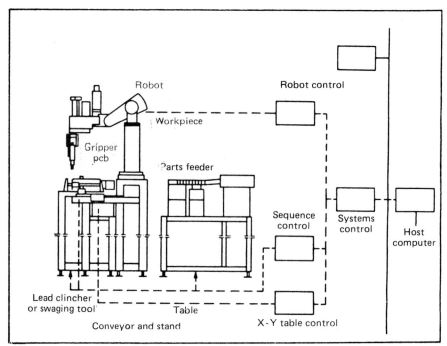

Elements of a robot assembly cell — the robot, parts feeders, conveyor and
controllers

Figure 4.4 Major Components of an Industrial Robot

Major Axes

- Rotation is the turning about the vertical axis that produces side-to-side motion.
- Vertical movement is the raising and lowering of the arm. In cylinder robots, this is accomplished by vertical linear motion of the arm. In polar robots, it is accomplished by rotation of the body about its horizontal axis.
- Radial extension is the movement of the arm in and out in relation to the body.

Minor Axes

- Wrist bend is the up-and-down motion of the hand caused by vertical rotation about a horizontal axis.
- Wrist yaw is the right or left motion of the hand caused by horizontal rotation about a vertical axis.
- Wrist roll is the twisting of the wrist.

    Some robots have the three major axes combined with only one or two of the minor axes. Robots are capable of point-to-point, or continuous path, operation. Most robots used in pick-and-place operations are point-to-point machines. Continuous path robots are used in operations such as painting and welding.

Fundamentals of Product Processes and Operations

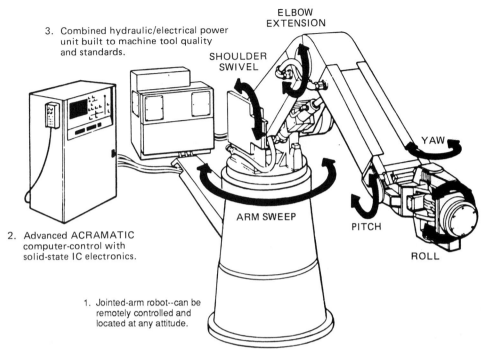

3. Combined hydraulic/electrical power unit built to machine tool quality and standards.

ELBOW EXTENSION

SHOULDER SWIVEL

YAW

ARM SWEEP

PITCH

ROLL

2. Advanced ACRAMATIC computer-control with solid-state IC electronics.

1. Jointed-arm robot--can be remotely controlled and located at any attitude.

Figure 4.5 A Six-Axes Robot (Cincinnati Milacron)

Another method of classifying robots is according to the manipulator's work envelope. A work envelope is the total area that the end of the robot's arm can reach. Three popular classifications are illustrated in Figure 4.6. Robots move according to the three illustrated designs. The design emulates man's physical capabilities by simulating his movements and activities through degrees of freedom that correspond to man's waist, wrist, elbow, shoulder, and fingers. Each type of robot has its own unique work envelope.

The Power Source. The manipulator needs power to drive it. Power for the manipulator can be developed from a pneumatic, hydraulic, or electric power source. Robots may therefore be classified according to their power source.

A *pneumatic power source* provides the compressed air to move the robot's manipulator through its work envelope. A major advantage of a pneumatic power source is its simplicity. The compressed air lines in the shop can be tapped for this source of power. A major disadvantage of a pneumatic power robot is its lack of accuracy in repeating tasks.

A *hydraulic power source* provides fluid power to drive the manipulator. A major advantage of using this power source is its load-carrying capability. Hydraulic robots have been constructed to lift in excess of 30,000 pounds. Another advantage is that a hydraulic robot does not generate sparks as it operates and thus can be used in a volatile environment—for example, in a spray paint booth. Some disadvantages

Computer-Aided Production Systems

117

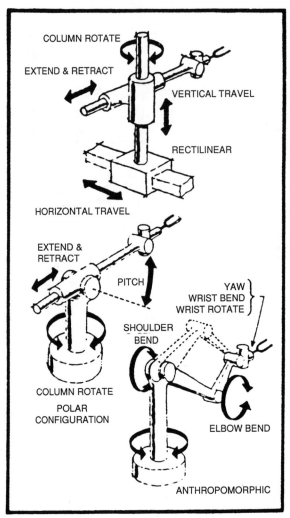

Figure 4.6 Classification of Robots By Work Envelope

of hydraulic-powered robots are the heavy maintenance requirements, the possibility of leaking, and the warm-up period required.

An *electric power source* for manipulators has become very popular because of its simplicity and easy maintenance. A major disadvantage of an electric robot is its low load-carrying capability. Future improvements in electric motor technology will overcome this handicap.

The Controller. The heart of a robotic system is the controller, which consists of a group of devices that govern the power delivered to the manipulators. Typical functions of a controller in a robotic system are to provide actions such as starting/stopping; accelerating/decelerating; regulating/protecting; and motions such as forwarding/reversing; upping/downing; and twisting/turning. A controller stores preprogrammed information for later recall, controls peripheral devices, and communicates with other computers in the manufacturing facility.

The interrelationship of the controller to other major components is shown in Figure 4.5. Controllers in the majority of current robotic systems are operated by 8-bit, 16-bit, or 32-bit microprocessors. A microprocessor-based controller allows it to be very flexible in its operation.

Robots may also be classified by the way in which the power supply sources to the robotic system are controlled: nonservo control robots and servo control robots.

Nonservo control. The nonservo control, often called an open-loop system, is the simpler of the two controls. It is called a "bang-bang" robot. That is, the controller sends a signal to the power source, which turns it fully on or fully off. In response, the manipulator moves until it runs into a stop (hence, the term bang-bang) placed in the travel path. There are no other points at which the parts of the manipulator can be stopped with any degree of certainty. As a result, each time a new movement of travel is used, the stops must be manually changed.

Servo control. The servo control, often called a closed-loop system, robot is more versatile than its nonservo counterpart. This control unit is capable of detecting the instantaneous position of the joint and can be programmed to stop between the end stops. A basic difference between the servo robot and the nonservo robot is a feedback system that allows the mechanics of the robot to communicate with the electronics of the controller to determine if an "error" exists between where the manipulator *is* and where it *should be.* If the controller recognizes that an error exists, it sends a signal to a motor to move the manipulator to correct the error.

Servo-controlled robots typically provide the capability to execute smooth motions with controlled speeds and sometimes even controlled acceleration and deceleration, which results in controlled movement of heavy loads. Servo-controlled robots can be programmed to any position on each axis anywhere in the work envelope with an accuracy of the end-of-arm positioning of the order of 1.5 millimeters (0.050 inch). These machines are more complex, and therefore they require sophisticated maintenance and are less reliable and more expensive than the nonservo robots.

The end effector. The end effector, sometimes called a gripping device or hand, is attached to the arm. The end effector is usually not considered to be one of the degrees of freedom. However, an end effector gives the robotic system the flexibility necessary for operation of the robot.

An end effector is generally designed to meet the needs of a particular application. As a result, end effectors vary from one application to another. These devices can be manufactured by the robot manufacturer or by the owner of the robotic system.

Most robots operate with an open-loop program, a kind of program that causes the robot to repeat the same set of motions continuously without modification. There are four principle ways of programming a robot: (1) manual method, (2) walkthrough method, (3) leadthrough method, and (4) computer terminal programming.

The manual method. Depending on the type of memory devices the robot has, they are variously programmed. Some robots operate by electromechanical devices such as switches that require some manual intervention.

Electromechanical devices require physical control devices, the programming of which requires preparing punched cards or tape and manually inserting the program into the device. Other robots use a magnetic tape, disk storage, or even microcomputers.

The walkthrough method. The walkthrough method requires the programmer to teach the robot by manually moving the manipulator through the desired sequence of positions. At each move, it signals the memory to record the position for playback during the work cycle.

The leadthrough method. In the leadthrough method of programming a robot, the programmer drives the manipulator through a sequence of positions by using a teach pendant, or console (Figure 4.7). The manipulator is lead to a desired position by pushing buttons on the teach unit. When the arm is finally at the desired position, the programmer pushes the RECORD or LEARN button to enter into the controller's memory this position of the manipulator's arm.

Figure 4.7 A Leadthrough Method of Programming a Robot

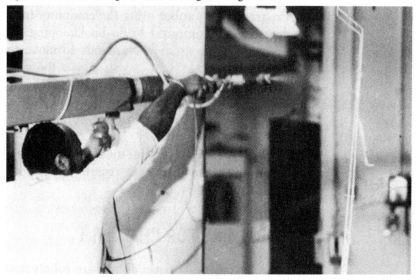

The computer terminal method.   Special languages with special commands are used to move the robot's manipulator during a programming exercise. Such languages allow the programmer to make the robot move more precisely than does either the walkthrough or leadthrough method. Computer terminal programming is not as popular as these two methods, however. In some cases, one of the popular methods is first used to teach the robot, and then the program is modified by using the computer terminal to make the robot perform precisely. As robotic technology advances, the computer terminal programming method will become more popular.

Uses of Industrial Robots

An industrial robot can be programmed to perform a variety of production tasks. Robots can perform two basic kinds of work in the factory: value-added work and nonvalue-added work.

Value-added work.   Value-added work is any task that increases the value of the part—for example, welding, spray painting, buffing, grinding, drilling, and assembly. In work of this kind, a tool is mounted at the end effector of the robot to perform certain jobs on the part to increase its value.

Because of the high degree of skill required in many welding applications, arc welding is a difficult application for certain robots. To achieve a good weld, many systems have been developed to allow the robot to sense the location of the seam to be welded and to modify its program. One method is to use the "through the arc" sensing system to monitor the electric current being drawn while the robot is welding. By monitoring the current being drawn, the robot can sense the width of the seam and adjust the welding speed to allow the gap to fill. Other systems use a camera mounted on the robot's arm. The camera looks at the seam and adjusts the robot's program accordingly. An application of a robotic welding system is shown in Figure 4.8.

Another application of robotic technology in assembly is spot welding, which is used to weld together thin sheets of metal. The spot welding guns are heavy, and a human operator soon becomes tired from moving the welding gun from point to point. The fatigue causes reduced productivity. Today, most spot welding on American-made cars is done by robots, and improved productivity is achieved. A spot welding activity is illustrated in Figure 4.9.

A spray booth is considered to be a dangerous place to work. The paint particles can cause serious health problems. Robots were quickly adopted for spray painting applications to replace human operators. Spray painting using robots presented some difficult problems that had to be overcome, two of which were maintaining the viscosity of the paint being sprayed and accurately locating the parts in front of the robot. These problems have been overcome in many applications, and spray painting is now a common application for robots in industry.

Nonvalue-added work.   Nonvalue-added work is also performed by robots. One example is transferring parts from one conveyer to another conveyer. Although transferring parts must be done in a manufacturing cell, the value of the part is not increased by the transferring operation. Other examples of nonvalue-added jobs

Figure 4.8 Robot Performs Arc Welding

performed by robots include material handling, machine loading, die casting, plastic injection molding, and forging.

Because moving materials around in a manufacturing plant does not add value to a product, industry has worked hard to minimize the costs of materials handling. Programmable controllers are used with a conveyer to allow flexibility in moving material in and out of storage and between the machines that perform the value-added work.

Robots are also used for "depalletizing"—the job of removing the material from the pallet and placing it on a conveyer. The parts or product can be "palle-

Fundamentals of Product Processes and Operations

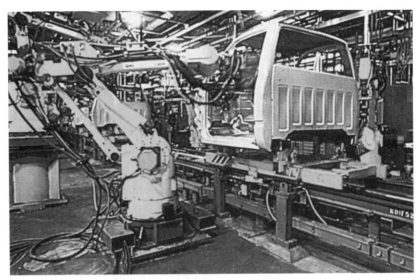

Figure 4.9 Robot Performs Spot Welding

tized" for shipment by robots after the product has been completed. The robot can pick up a part from a conveyer, load the part into the machine tool, send a signal to the machine to clamp the part, pull clear of the machine, and send a signal to begin the machine cycle. When the machine cycle is completed, the robot unloads the part and reloads the machine with another part to be machined.

Numerical Control

The program of instruction component of an NC system was discussed in Chapter 3. The machine controller and the machine tool components are discussed here and are illustrated in Figure 4.10. This mill and drill machine can be operated by numerical control. On an NC machine, the motors are controlled by the machine, not by hand wheels.

Traditionally, when the NC method of machining a part is used, the part drawing is studied by the programmer who then converts the information on the drawing to the necessary code. The code represents every movement, path, or action the machine tool must take to properly machine the part as described by the engineering drawing. The complete series of codes necessary to produce the part is called a program. The programmer's function is to write part programs that efficiently produce parts called for by engineering drawings.

In the most basic NC system, the programmed instructions are stored on punched tapes and interpreted by electromechanical tape readers connected to the machine tool. More advanced systems use computer numerical control in which the machine is controlled by a dedicated microprocessor.

The program instructions control complex machining operations that create highly reproducible tolerances on widely available equipment with minimal human intervention. The automation of entire assembly operations has already begun

Figure 4.10 Factory Floor Components of an NC System

through the increasing use of robots in certain industries. However, there is still too much human involvement with set-up time, raw stock selection, feeding, and materials handling. In addition, most automation efforts have been rigid and intended for mass production. The need to overcome these problems has prompted the automation industry to develop more sophisticated systems.

Today, the most sophisticated systems use distributed numerical control when the machine is controlled by a network of computers and is linked to robots, programmable controllers, and inspection systems, although distributed numerical control, part programs, and other data can be directly downloaded or uploaded in CNC systems. Such systems can also be used with robots and other computer controls to create manufacturing cells to form the basis for a flexible manufacturing system (FMS).

Adaptive Control

Some CNC systems incorporate in-process gauges or tool breakage detectors into the machine tool. These machines are equipped with adaptive control systems (AC) that measure the spindle deflection or force, torque, cutting temperature, vibration, and horsepower. They use these process variables to modify the machine speeds and feeds. The benefits gained are the increased production rates and increased tool life and lower machining costs.

Materials Handling System

The materials-handling system in a computerized production environment serves two functions. The first is to move parts between machines (primary system). The second is to locate the parts on machines for processing (secondary system).

Automated Guided Vehicles

An automated guided vehicle system (AGVS) (primary system) can be used to transport parts and materials between machine lines and storage areas, depending on the complexity of the AGV system. The central computer downloads commands to a floor-level computer, which in turn translates general commands into specific dispatching instructions to the AGVs. The AGVs are best suited for situations where there is some need for random access between two or more pairs of pickup and drop-off points.

Some vehicle designs use wire-guided navigation systems. The vehicle tracks small floor-embedded wires that generate magnetic fields. Newer optically guided navigation systems use invisible ultraviolet stripes painted on the floor.

The secondary parts-handling system presents parts to the individual machine tools. The secondary system generally consists of one transport mechanism for each machine that interfaces with the primary handling system, locating parts at each machine for processing.

Robots in Materials Handling

Robots play an important role in the computerized materials-handling system. Two typical functions robots perform are material transfer and machine loading.

Materials transfer. In materials transfer, robots are used to move workparts from one location to another (Figure 4.11). Examples of these applications include pick-and-place, transfer of workparts from one conveyer to another conveyer, stacking, loading parts from a conveyer onto a pallet in a required pattern and sequence, and loading parts from a conveyer into boxes.

Machine loading. In machine loading, the robot works directly with the processing machine to load or unload parts of materials from the machine (Figure 4.12). In upsetting and stamping operations, the robot holds the workpart while it is being processed by the machine.

Transport systems. Transport systems are constructed with various sizes and lengths of strut, which fit together to form operating stations and transport lines. A transport system is illustrated in Figure 4.13. Modules such as drives returns, transverse conveyers, lifts and turns, stop gates, proximity switches, and pallets are used to transport products from one workstation to another under computer control.

Figure 4.11 Robot Performs Material Transfer

Fundamentals of Product Processes and Operations

Figure 4.12 Robot Performs Machine Loading

Automatic Storage/Retrieval System (AS/RS)

Six major processes exist in most manufacturing facilities, as follows:

1. Receiving
2. Inspection
3. Picking
4. Production
5. Assembly
6. Shipping

An automatic storage/retrieval system (AS/RS) is a key element in a computer-aided materials-handling system that serves as a common link between manufacturing operations and storage areas. From the time material is received until the finished product is shipped, it may be picked up, moved, stored, worked on, and handled dozens of times. While the material is going through the production cycle, it must be tracked and controlled in real time for positive inventory control. An AS/RS is shown is Figure 4.14.

Receiving

In an automated receiving area, material is managed and controlled from the moment it arrives on the dock. It is quickly identified, and data about it are input into the inventory database by means of a keyboard terminal, light pen, magnetic wand, or other automatic identification device.

Figure 4.13 Product Transport System

The computer that is linked to the factory's material requirement planning (MRP) database verifies the identification and quantity of the incoming material against material orders. The computer then prints a move ticket that is attached to the material and that directs the material to storage, inspection, order picking, manufacturing, assembly, or other activities.

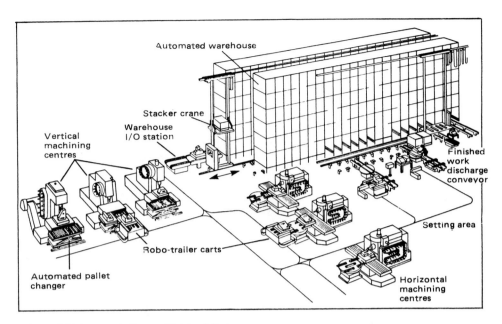

Figure 4.14 Automatic Storage/Retrieval System

The material is then moved out of the receiving area by an automated transportation system such as a conveyer, in-floor towline, or AGV. Materials destined for storage move to the automated high-rise storage system. A storage/retrieval (S/R) machine picks up the load, stores it in the preassigned storage aisle and storage rack opening, and verifies to the computer that the load has been correctly stored.

Inspection

A computer-aided quality control system determines the size of the lot samples to be sent to quality inspection. The material then can be sampled, sent to quality control, and the remaining shipment moved to storage. The material will be guaranteed in the computer and cannot be accessed until released by quality control.

Picking

An automated storage/retrieval system can reduce the time required to pick materials by bringing the material to the picker because picking is done at a well-lighted workstation. Picking rates have been increased and error reduced. When a pick is completed, the picker signals the system computer and the computer automatically updates the inventory records in real time, thus providing the management with timely information. An AS/RS is shown in Figure 4.15.

Figure 4.15 Picking Area in an Automated Storage and Retrieval System

Production

To assure that machine tools and equipment have a high operating uptime, traditional production practice usually stores additional materials at each workstation and between manufacturing operations. This work-in-process inventory often exceeds the space in which the associated productive work is performed.

With an automated AS/RS combined with automatic identification systems and automatic transportation systems, such as automatic conveyers or AGVs, material can be quickly and accurately delivered to the workstations for processing.

Assembly

Storing parts, materials, and assemblies in an AS/RS reduces the amount of material needed at a given time along the assembly line. The unplanned shutdown of an assembly line due to slow delivery of materials can be avoided.

Shipping

In an AS/RS system, materials and finished products move from assembly line through automatic packaging and automatic palletizing into the shipping/warehouse area where they are identified to the computer. According to shipping schedules, they are sent to the shipping dock or storage in the high-density AS/RS system. The speed and accuracy of an AS/RS system reduce floor space requirements for staging orders before shipment (Figure 4.16).

Production Planning and Control

Traditionally, the authority-to-make document activates manufacturing engineering operations and production planning and controls operations (see Chapter 1). Production planning and production controls are the two functions that drive fabrication, assembly, and test and inspection operations on the factory floor.

Production Planning

The authority-to-make document contains information in the form of instructions to the performing and supporting departments, and copies are distributed to each responsible function, along with all the released engineering drawings

Figure 4.16 Automated Storage/Retrieval System

and specifications necessary to define and establish the configuration database for each function.

The forecast delivery schedule transmitted by the MA document is converted into a master schedule by the production planning department. A master schedule, sometimes called a production plan, generally specifies a broad schedule of the total number of units to be produced, but it does not specify the details of accessories and specific products. That is, it may be viewed as a high-level schedule from which detailed schedules are made. Typical detailed schedules include material requirement planning, capacity planning (see Chapter 6), testing and inspection planning, and resource planning such as are needed for workers, machines, and facilities. These schedules are developed around the just-in-time (JIT) concept (Chapter 6), which is a philosophy of scheduling production resources so that they arrive at the right location at the exact time for production operations to be carried out according to schedule.

Production Controls

The production control function is responsible for directing or regulating the movement of goods through the entire production cycle (see Figure 1.7), from the requisitioning of raw materials to the delivery of the finished product. It is the nervous system of the production function that sends directives to the factory floor. These directives outline the individual operations on all component parts and tell how to put them together into finished products. A factory must have production controls to perform the many operations required to make parts and to produce the finished products at the required time. Production control is an ongoing activity designed to strike a balance between several conflicting objectives, or goals.

Many controlled operations that may be classified in many ways take place in the production function. For discussion, the production control work is listed under nine headings as follows:

1. Quality control
2. Data collection
3. Customer order entry
4. Materials requirements planning
5. Capacity requirements planning
6. Shop-floor control
7. Purchase order
8. Inventory controls
9. Production cost accounting

Various automated systems are in use today to assist the production control process. Several typical systems are discussed in the following several paragraphs.

Problems that occur during the production cycle of activities include poor quality product, misrouting of parts, improper inventory stock, poor tracking of products, poor utilization of machines and equipment, and improper costing of a product. Many of these problems result from the inability of the traditional approach to deal with the complex and changing nature of manufacturing. Consequently, production falls behind the planned schedule because of lack of labor or equipment, or raw materials inventories may become too high for work in progress or for finished products. In addition, poor production scheduling can result in job interruptions, lack of current bills of materials, and inaccurate inventory records, all of which result in decrease in productivity.

With the development of computer-integrated production management (CIPM) systems, however, management is able to gain a comprehensive range of management control over the various production functions because CIPM is a philosophy of integrating the production control systems through a common data link. In this text, production planning and control functions are considered to be business components. The integration of these functions into the manufacturing environment can be viewed as computer-aided business (CAB). Computer-aided business functions are discussed in Chapter 6.

Computer-Aided Quality Control

The objectives of computer-aided quality control (CAQC) are to improve quality and increase productivity. The key to CAQC is to automate the testing and inspection processes through the application of computers with advanced sensor technology. Traditionally, quality control (QC) has been performed using manual inspection and statistical sampling. In CACQ, there is inspection of every individual finished piece rather than of random samples. Computer-aided inspection and computer-aided testing are subsets of CACQ.

Computer-aided inspection. Inspection during production can be integrated into the manufacturing process rather than requiring that the parts be taken to an inspection area. It can be done using both contact and noncontact inspection methods.

Contact inspection methods. Contact inspection methods usually involve the use of coordinate measuring machines (CMM). Figure 4.17 is an example of a CMM. In this method of testing, parts must be stopped or repositioned to make physical contact with the CMM.

Noncontact inspection methods. Noncontact inspection is obviously much faster than contact inspection. Most noncontact inspection methods use optical sensing techniques such as machine vision, scanning laser beam devices, or photogrammetry. Machine vision technology is being advanced continually (Figure 4.18).

Figure 4.17 Coordinate Measuring Machine

Future machine vision systems will have better image resolution, greater ability to distinguish gray areas and color, and more intelligence and memory for improved object recognition capability than current systems have.

Computer-aided testing.   Computer-aided testing is simply the application of the computer to the testing procedures. Such systems are becoming a commonplace in many manufacturing enterprises.

Computers can be used to monitor the test, analyze the results, and prepare a report of the results. At a much higher level of automation, CAT cells are integrated directly into production lines. During operation, a product such as an automobile engine is transferred by a materials-handling system to a test station. The computer monitors the test, collects the data, and analyzes the results. If the product fails the test, it is then transferred to final parking or to another station for further diagnosis.

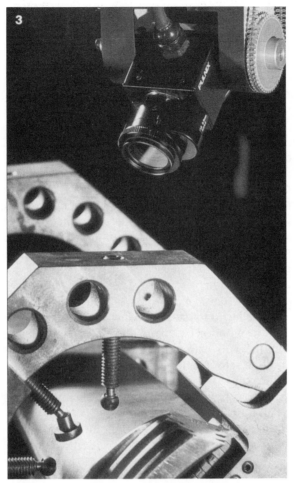

Figure 4.18 Machine Vision System

Computer-aided testing cells are typically applied in situations where the product is complicated and produced in large quantities. Examples include electronic integrated circuits and automobile engines.

Factory Data-Collection System (FDCS)

The purpose of a factory data-collection system in a shop floor control (SFC) environment is to provide the basic data for monitoring the progress of an order. The system consists of input devices that are located at various points where data are created and that bring data from these points to a central point.

The system collects such data as the name of the operator in charge of the process, process machine information, parts information, parts counts, rejection rate, and labor time spent on a particular job. Workers' time and attendance data are also collected for use by payroll and accounting systems.

Traditional factory data collection uses job travelers (reports attached to the product as it moves through the production cycle), employee time sheets, or operation tear sheets to track order progress. These methods involve a time lag, so that the order status report may not be prepared until the end of the work shift.

In the collection of factory data based on real time, simple data input terminals, such as those shown in Figure 4.19, are located throughout the factory floor. Workers insert precoded cards or badges at the terminal or at the workstations. The terminals collect information to identify the worker, the job and its location, the workstation, the machine, and the start and stop times. This information is transmitted directly to the host computer.

Data collection terminals at workstations are usually low-cost terminals that can be configured to handle input from a range of standard devices (Figure 4.19). They include bar-code readers, card readers, badge readers, magnetic-stripe reader, and keyboards equipped with user-defined function keys. Portable terminals are also used for stocktaking, shipping, or other mobile applications (Figure 4.20). The system control unit provides the hardware interface between the data collection terminals and the host computer.

Computer process monitoring. Computer process monitoring is a special kind of data-collection system in which the computer is connected directly to the workstation to monitor order progress. The plant manager can analyze data collected from the workstation and make changes in the plant operations.

Condition sensors and position monitoring sensors are widely used in monitoring applications. Condition sensors are used to sense temperature in current, vibration, airflow, pressure, humidity, and chemical flow. Position monitoring sensors are used for counting and timing.

Current sensors are used to detect current draw on motors to predict either motor failure or some other consequence. A current sensor can be used to help elim-

Figure 4.19 Data Input Terminals

Figure 4.20 Portable Terminals for Mobile Applications

inate costly motor repairs by slowing a machine down. By detecting the current drawn by a motor, current sensors can also be used to indirectly sense variables other than current. For example, a current sensor can function as a presence sensor in a conveyer system by determining the difference in current draw on the motor when the conveyer belt is full rather than empty.

Electromechanical limit switches, proximity sensors, and photoelectric sensors are used in monitoring production. These sensors can be used to collect production data, such as the number of good parts versus bad parts, the amount of time a machine is up or down, and cycle time. Real-time access to the information gives the manager in charge of production control the data needed to monitor the production system.

Customer Order Entry and Control

The entry and control of customers' orders are used to capture information from the sales department (see also Chapter 6). These functions typically include tracking customer orders, and reporting sales analysis. A typical customer order report is illustrated in Figure 4.21.

Material Requirements Planning

The material requirements planning (MRP) function is a set of processes that use the bill of materials, on-hand and on-order inventory data, and the master plan to determine when to order raw materials and components for assembled products (see also Chapter 6). It can also be used to reschedule orders in response to changing production priorities and demand conditions.

Capacity Requirements Planning

Capacity requirements planning (CRP) is needed to establish, measure, and adjust the labor and equipment resources needed to meet the production schedule (Chapter 6). It reports the labor hours required for each work center for both released and unreleased work orders. The report also shows the capacity of each work center for comparison with the load. A CRP report is shown in Figure 4.22.

Figure 4.21 Customer Order Report

Work Order Status ORDER STATUS

Order Overview | Chg Lot Status | Stop | Printed Report | Both | Exit

Order # W017658 | Part Number 15456 - 18955 | Discription REAR PANEL FRAME

Prtial Number	Work Center	Work Station	Oprtn	--In-- Date	--In-- Time	--Scheduled-- Beg Date	--Scheduled-- End Date	Qty In	Qty Bonus	Stat	Prtial Parent
	METAL	NODRILL	DRL1	06/15	03:36P	06/15	06/16	50	0	5000	
	METAL	DEBURR	DB04	06/16	08:10A	06/16	06/19	50		5000	
	METAL	PAINT	PT10	06/19	08:30A	06/19		49		500	

Current location of work order

Quantities moved in from previous sequences

Quantities added at each sequence

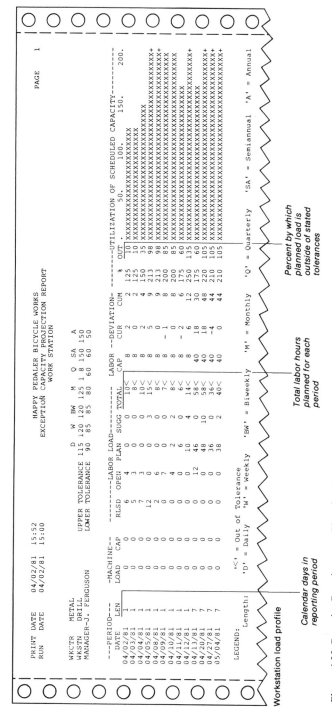

Figure 4.22 Capacity Requirements Planning Report

Shop Floor Control

The SFC function schedules and dispatches work to the production process, tracks the usage of material, labor, and machines, and records unit dispositions and the associated quality data. Within this function, a complete record of the manufacturing process is maintained. A typical report is illustrated in Figure 4.23. Further discussion of SFC is in Chapter 6.

Purchase Order Entry and Control

The entry and control of purchase orders requires the entering, maintaining, and tracking of all purchase orders. The order-need dates are kept up to date by MRP for the purpose of priority planning. A purchase order tracking report is illustrated in Figure 4.24.

Inventory Control

The inventory control module of the computer-aided production planning and control system has two major functions:

1. Inventory accounting
2. Inventory tracking and control

Inventory accounting is concerned with inventory transactions and records. A typical report is shown in Figure 4.25. Inventory tracking and control are concerned

Figure 4.23 Product Scheduling Report

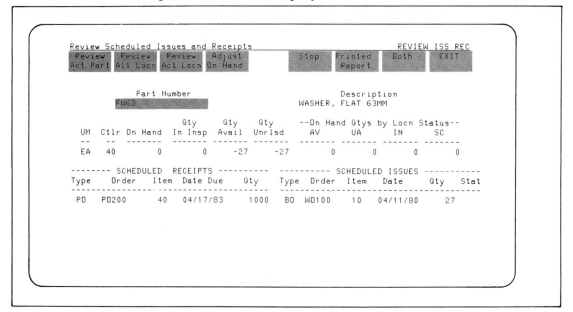

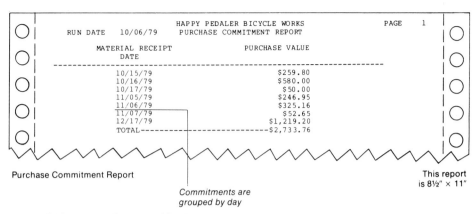

```
                      HAPPY PEDALER BICYCLE WORKS              PAGE    1
  RUN DATE    10/06/79        PURCHASE COMMITMENT REPORT

      MATERIAL RECEIPT                    PURCHASE VALUE
           DATE
  -----------------------------------------------------------------
           10/15/79                          $259.80
           10/16/79                          $580.00
           10/17/79                          $ 50.00
           11/05/79                          $246.95
           11/06/79                          $325.16
           11/07/79                          $ 52.65
           12/17/79                        $1,219.20
           TOTAL-------------------------$2,733.76
```

Purchase Commitment Report

This report
is 8½″ × 11″

Commitments are
grouped by day

Figure 4.24 Purchase Order Tracking Report

with economic lot size, safety stock levels, and inventory analysis. They also determine reorder points (Figure 4.26).

Production Cost Accounting

Production cost accounting deals with the expected costs and the actual costs of manufacturing a product, and the differences between actual costs and expected costs. A product costing report is illustrated in Figure 4.27.

The expected costs of a product are compiled on the basis of the following sources of accounting data:

1. Overhead rates
2. Material costs
3. Bill of materials
4. Route sheets
5. Time standards
6. Labor and machine rates

The actual cost of a product can be calculated on the basis of the actual material costs and labor costs. Overhead costs are usually excluded because they do not represent an actual expense of the product but, rather, an allocation of general factory and corporate expenses. The cost accounting software also compares standard costs to actual costs and generates variance reports on labor, materials, and purchases.

Summary of Production Operations Control

The master schedule stored in the computer database serves as the primary source of all planning for material, equipment, machines, and human resources. The entry of the product configuration data into the production control system

```
CONTROLLER   20                    HAPPY PEDALER BICYCLE WORKS                                    PAGE   1
RUN DATE   11/12/79                  INVENTORY VALUE REPORT
```

PART-NUMBER	DESCRIPTION	PART CLASS	ON-HAND QTY	ON-HAND VALUE	IN-INSP QTY	IN-INSP VALUE	SIX-MO RQMT QTY	SIX-MO RQMT VALUE	CUM% TOTAL	ABC	DAYS SUPPLY
200202	CENTER PULL BRAKE ASSEMBLY	F	1,600	20,800	0	0	3,325	43,225	62.9	A	50
200201	CABLE CARRIER ASSEMBLY	F	166	996	0	0	3,325	19,950	91.9	D	5
200504	SPROCKET 17T	P	100	110	100	110	1,880	2,068	94.9	D	5
100016	REAP BRAKE ASSEMBLY	F	5	75	0	0	1,023	1,823	97.5	D	1
200700	BODY ASSEMBLY	F	0	0	0	0	1,593	1,593	99.9	D	0
200505	SPACER	P	0	0	0	0	1,880	940	100.0	D	0
100070	AIR BAG	F	0	0	0	0	0	0	100.0	D	0
200500	SNAP RING	P	0	0	0	0	0	0	100.0	D	0
200501	TOP PROTECTOR	P	0	0	0	0	0	0	100.0	D	0
100000	FRAME ASSEMBLY	F	0	0	0	0	0	0	100.0	D	0
******************************** TOTALS:				$21,981		$110		$68,753			

Valued at
standard cost

Cumulative
percent of
total six month
requirement

This report shows the value of on-hand inventory and the total six month requirement of inventory for each controller. It provides a complete snapshot of present conditions and forecast requirements.

Figure 4.25 Inventory Accounting Report

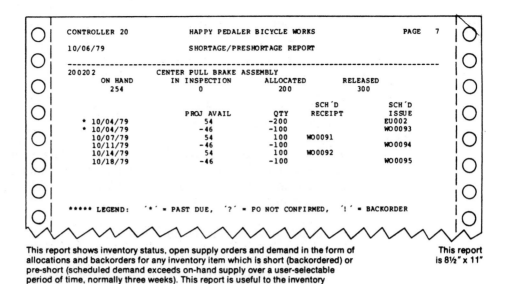

```
CONTROLLER 20              HAPPY PEDALER BICYCLE WORKS                    PAGE    7

10/06/79                  SHORTAGE/PRESHORTAGE REPORT
-----------------------------------------------------------------------------------
20 0 20 2                  CENTER PULL BRAKE ASSEMBLY
         ON HAND           IN INSPECTION        ALLOCATED           RELEASED
           254                  0                 200                 300

                                                    SCH´D              SCH´D
                          PROJ AVAIL       QTY     RECEIPT            ISSUE
   *  10/04/79                54          -200                        EU002
   *  10/04/79               -46          -100                        WO0093
      10/07/79                54           100      WO0091
      10/11/79               -46          -100                        WO0094
      10/14/79                54           100      WO0092
      10/18/79               -46          -100                        WO0095


  ***** LEGEND:    ´*´ = PAST DUE,   ´?´ = PO NOT CONFIRMED,   ´!´ = BACKORDER
```

This report shows inventory status, open supply orders and demand in the form of allocations and backorders for any inventory item which is short (backordered) or pre-short (scheduled demand exceeds on-hand supply over a user-selectable period of time, normally three weeks). This report is useful to the inventory controller because it gives an early warning about problems and potential problems while there is time to expedite and avoid shortages.

This report is 8½" x 11"

Figure 4.26 Inventory Tracking and Control

(PCS) is initiated by manufacturing engineering. The configuration of the product is defined in the released engineering drawings and the referenced specifications.

Many functional operations take place during these phases, as well as subsequent production phases. Typical of these functional operations are the following:

- Maintenance of product configuration
- Generation of production requirements

Figure 4.27 Product Costing Report

```
                         HAPPY PEDALER BICYCLE WORKS                    PAGE  1

  11/02/79                  COST SHEET FOR: 100002        HEAD STEERING KIT   LL=01

                                                          CHG%=25%

                       STANDARD           CURRENT
                       --------           -------
  LABOR        -TL      0.2600            0.2654
              -LL      0.5300            0.5337
  MATERIAL             0.5500            0.9592*
  OVERHEAD    -TL      0.0600            0.0581
              -LL      0.1300            0.1177
                      ---------          ----------
  UNIT COST            1.5300            1.9341*       Exceeds selected
                                                       allowable difference
                                                       between standard and
                                                       current costs
```

Product Cost Sheet

This report is 8½" × 11"

- Preparation for parts fabrication
- Fabricated parts production
- Procurement of purchased parts
- Release of material to assembly
- Selection or design of test and inspection equipment
- Material processing and assembly

These functional operations and their detailed production operations must be managed. The end objective of the production control system is to coordinate and manage the many functions and responsibilities of the total organization in satisfying three demands:

- A quality product must be produced.
- The product must be delivered on schedule.
- The product must be produced within budget limitations.

In facing these demands, the CIPM system is constantly being reviewed, modified, expanded, and improved as required to meet the ever-increasing pressures of technological change, economic conditions, and customer requirements. With the power and flexibility of a well-designed CIPM system, the many facets of the production process are kept in balance and on schedule through a structure made up of management information systems (MIS), PCS, and timely management control reports.

Computer-Aided Manufacturing

Computer-aided manufacturing (CAM) carries the broadest implication of all the terms in the production function (Figure 4.28). It is the technique of using a computer and supporting processing software to aid in the production of industrial goods or provide factory floor information (Figure 4.29). It embraces the implementation of the design, the processes, the machines, and materials-handling system to actually produce a specific product that meets all design requirements.

A CAM Scenario

Once the design is complete [1, 12], the feasibility of producing it economically is verified by a manufacturing engineer. Because the design has been stored in the database, the engineer can easily gain access to it. In fact, because of the ease of sharing data, the feasibility of manufacturing a part may be verified during design. In this fashion, the computer facilitates the ability of various people to work together, to reduce costs, and to increase productivity.

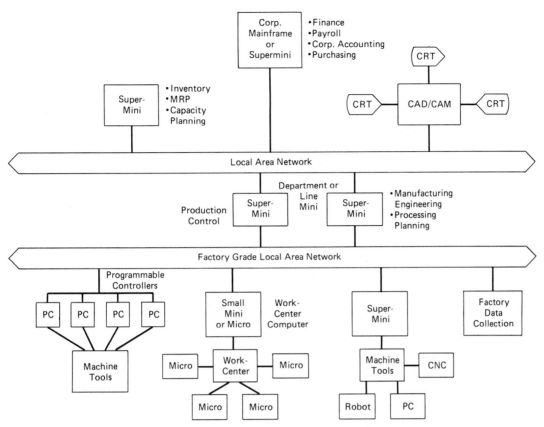

Figure 4.28 Applications of Computers in CAM

Figure 4.29 CAM Systems in Production

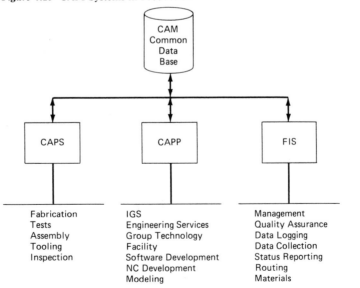

CAPS	CAPP	FIS
Fabrication	IGS	Management
Tests	Engineering Services	Quality Assurance
Assembly	Group Technology	Data Logging
Tooling	Facility	Data Collection
Inspection	Software Development	Status Reporting
	NC Development	Routing
	Modeling	Materials

Computer-Aided Production Systems

Tooling. The tooling to build the system is selected by using the database to reduce the effort of selecting and designing suitable production tooling. Again, the computer searches the designs that can be modified easily.

If a high-volume, parts-production system is being designed, its performance is simulated. The simulation provides performance data that permit the customer to evaluate the system. For those parts of the system that will be built on an NC machining system, information about the location of cutter paths is automatically generated by the computer. Because the tool path of the NC machine can be verified on a computer monitor, the cost of part programming is reduced, as is the number of bad parts produced during setup. In a similar manner, information is produced for the computer-controlled robots that assemble the system. The designer verifies the operation of the robots by observing simulated performances on a computer monitor.

Process and materials planning. Bills of materials, production schedules, process flow data, and other manufacturing information are then produced automatically by the computer for use on the shop floor. Much of the information needed is in the database at this time. If additional data are needed, they can be input by the operator at the terminal.

Shop floor monitoring and control. While the product is being built, the manufacturing process is continuously monitored by computers. The process is adjusted for efficiency, quality, and cost; and raw materials and completed parts are moved and stored by automatic materials-handling and warehousing systems. The cost data are also recorded in the data base so that they are available for any bidding on a similar component. Control information is generated from manufacturing data and information about the factory layout.

Management reports. Using the information in the manufacturing database, the computer generates periodic reports so that managing the factory environment is possible. Moreover, managers have access to the computer database for current manufacturing reports, which present data in statistically meaningful formats.

Exercises

Self-Study Test

1. What are the components of an NC system?

2. Explain the fundamental concepts and operating principles of NC.

3. Give five examples of applications of numerical control in automated manufacturing systems.

4. Explain the fundamental concepts of adaptive control.

5. Explain the difference between numerical control and robotics.

6. How are robots constructed and how do they work?

7. What are robots capable of doing in terms of accuracy, repeatability, speed, and load-carrying capacity?

8. Discuss the methods of programming a robot.

9. Explain the applications of industrial robots and examine the characteristics of potential applications that support the use of industrial robots.

10. Describe the materials handling and storage applications in automated manufacturing operations.

11. Explain the term flexible manufacturing system (FMS).

12. Discuss the principles of automated inspection and the important sensor technologies that are used to implement it.

13. List the various functions that constitute the information processing cycle that occurs in a manufacturing firm.

14. What are the principal roles of the computer in the information processing cycle in a manufacturing firm?

15. Explain the two principal planning activities, material requirements planning (MRP) and computer-aided process planning (CAPP).

References

[1] Chang, T. C., *An Introduction to Automated Process Planning Systems.* Prentice-Hall, Englewood Cliffs, NJ, 1984.

[2] Wolovich, W., *Robotics and Basic Analysis and Design.* Holt, Rinehart and Winston, New York, 1987.

[3] Groover, M. P., *Automation, Production Systems, and Computer-Integrated Manufacturing.* Prentice-Hall, Englewood Cliffs, NJ, 1987.

[4] Hoekstra, R., *Robotics and Automated Systems.* South-Western, Cincinnati, OH, 1986.

[5] Plossl, K. R., *Engineering for the Control of Manufacturing.* Prentice-Hall, Englewood Cliffs, NJ, 1987.

[6] Goetsch, D. L., *Fundamentals of CIM Technology.* Delmar, Albany, NY, 1988.

[7] Hordeski, M. F., *CAD/CAM Techniques.* Reston, Reston, VA, 1986.

[8] Miller, R., "Where CIM Is Trust Business as Usual." *Managing Automation,* August 1988, pp. 62–64.

[9] Klein, A., "Robot Arc Welder Intended for Low-Volume Applications." *Managing Automation,* June 1988, pp. 26–29.

5

Integration of CAD and CAM Technologies

Computer-Integrated Production

Advancements in computer technology continue to have an impact on the manufacturing process and enhance integration of the individual islands of automation in computer-aided design (CAD) and production (CAP). The term computer-aided production (CAP) is used interchangeably with computer-aided manufacturing (CAM). Even though there are many islands of automation in the production (CAD/CAM) cycle, there are still walls between the communication of engineering and production (manufacturing) functions. As a result, engineering continues to toss design information over the walls to production. Figure 5.1A illustrates engineering passing data to production *before* automation; Figure 5.1B illustrates the same process *after* automation. In many cases for both illustrations, production receives inaccurate data, or the data are in such shape that they cannot be properly interpreted by production. This communication problem causes production problems, time delays, and poor project quality.

Figure 5.1 also shows that there is very little, if any, sharing of computer data between the two major functions. Besides a lack of sharing, additional problems result from the walls between the two major functions. Typical of such problems are the following:

- Increased human errors
- Long programming time for numerical control
- Long turnaround time

Figure 5.1 Engineering Passing Data to Production

ENGINEERING/PRODUCTION INTERFACE WITHOUT CAD OR CAM

5.1A

ENGINEERING/PRODUCTION INTERFACE WITH CAD/CAM

5.1B

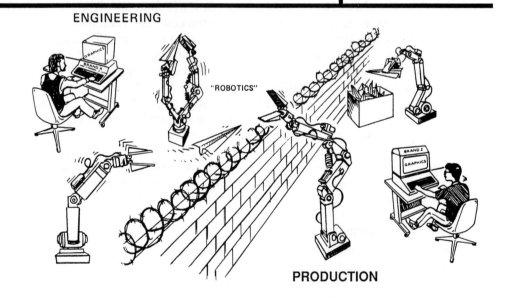

- Decreased productivity
- Reduced producibility information to engineering

Such problems have prompted engineering and production to study methods and implementation strategies for computer assistance in integrating certain subsystems of the two functions. Thus, computer integration of these two functions brings into focus automated computer assistance in sharing data between operations in the production cycle (Figure 1.7).

Through cooperative efforts between the two functions, computer integration of subsystems of CAD and CAM begin to take place. This integration of CAD and CAM into the production process to improve productivity is referred to as CAD/CAM. Such systems store, retrieve, manipulate, and display graphic information—all with unsurpassed speed and accuracy. Thus, more can be accomplished in a given time by engineers, who are an increasingly scarce and valuable resource. Product quality and yield are also improved, optimizing the use of energy, materials, and manufacturing personnel. New CAD/CAM technologies, including hardware and software concepts, appear continually in the marketplace as a logical result of ongoing evolutionary processes in human thought, industry, and striving for manufacturing improvements. The majority of these technological improvements usually represent only small advances that can be implemented with relative ease by careful planning, support from top management, and by simple addition, replacement, or modification of existing systems. Development and coordination of CAD/CAM technology are being subsidized by various agencies from private and public sectors of the industrialized world [1, 5].

The ICAM Program

The U.S. Air Force program for integrated computer-aided manufacturing (ICAM) [2, 1] was brought about by needs and pressures in state-of-the-art technologies, economics, increasing human limitations, aerospace design and manufacturing complexity, computer developments, and competition from abroad. The ICAM program is essentially a plan to produce systematically related modules for efficient manufacturing management and operations. As a primary goal, the ICAM program is a practical effort to greatly shorten the implementation timespan for incorporation of compatible and standardized techniques and to provide unified direction for industry. Some of the ICAM technology may be individually implemented in industry with short-term gains, but the primary benefits of the modular structure shown in Figure 5.2 are more evident in a fully integrated system.

As illustrated in Figure 5.2, the ICAM logo may serve as a planning and implementation guide for computer integration techniques of CAD and CAM. It supports all areas of CAD/CAM technology illustrated in Figure 5.3. Integrated CAD/CAM systems are capable of producing parts from a designer's specifications. That is, CAD/CAM includes all operations in the production cycle (Figure 1.7) that have been computerized and tied into a single controlling and coordinating system. The figure illustrates many applications of CAD and CAM, including management planning.

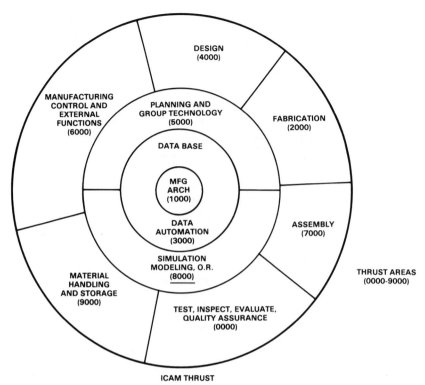

Figure 5.2 ICAM Program Thrusts

Refer again to Figure 5.2. The ICAM program logo is a graphic representation of the thrust areas. The concentric pattern captures the transition from the inner thrust areas of concepts out to the eventual real world of the shop floor and production line. A thrust area is the endwise push exerted from the inner ring to the outer ring in developing a concept. For program identification, each segment of the logo is labeled with a four-digit number. The logo also defines the structure and components of an ICAM program and their interrelationships.

CAD/CAM Integration

The ICAM program is a long-term effort that includes the establishment of modular subsystems, which are designed to computer assist and integrate various phase of the design, fabrication, and distribution processes, and their associated management hierarchy, according to a prioritized master plan. In essence, the ICAM program provides a basic roadmap for integrating CAD and CAM where computer technology is applied to the design and production of an item in pursuit of a profit.

Data communication among the various CAD/CAM subsystems is accomplished through a shared access to a common database system. The CAD/CAM subsystems are referred to as computer-integrated manufacturing subsystems

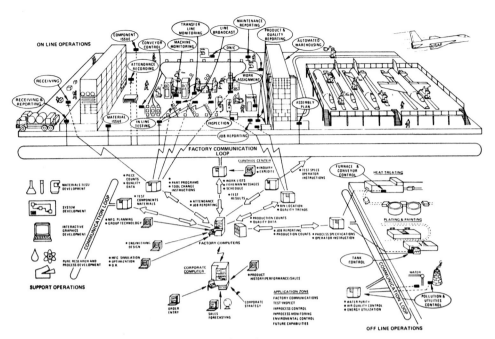

Figure 5.3 Integrated Computer-Aided Manufacturing

(CIMS). A CIMS has the specifications and characteristics that allow it to be computer integrated into a CIMS environment. Computer-Aided Manufacturing–International, Inc. (CAM–I) developed a system framework for the integration of manufacturing functions through computerization (see Figure 5.4). The framework shows the boundaries within which applications should be constrained, and the needed interfaces and interactions between them. The main objectives of the project are to do the following:

- Define a system framework for CAD/CAM that is as independent as possible of present or predicted computer hardware and software.
- Describe an overall CAD/CAM system framework capable of simple orientation to an individual company.
- Provide a means for easy and efficient replacement or enhancement of individual applications systems.
- Provide a means whereby users may involve their own pace into the total CAD/CAM system through the use of a common database system.
- Provide boundaries and guidelines for future development activities.

The model shown in Figure 5.4 illustrates a composite view in a standard hierarchical "tree" pattern of the thrust areas. Data communication between the thrust areas is also illustrated. The highest node in the hierarchical tree structure is product requirements. The other nodes are minor.

Integration of CAD and CAM Technologies

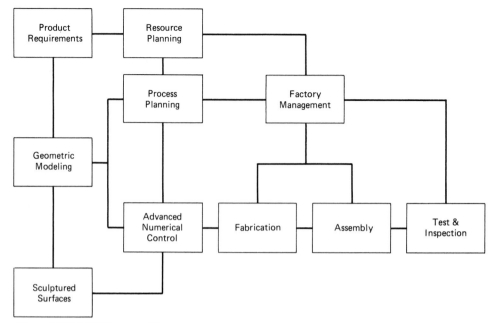

Figure 5.4 CAM–I Framework Project

The Big Three in CAD/CAM

At this point, you should review Figure 1.12. As shown there, one role of the Big Three (CAD, CAPP, and CAM) is to cogenerate a common data base for product design. This database provides common data to the production process and is often referred to as the engineering released database (ERD). Engineering releases the data of the product model to engineering data control (EDC) when the product model meets the desired design requirements. Engineering data controls all changes made in the data of the product after it has been released by engineering and also protects the integrity of the data.

Design, process planning, and producibility information are provided to the common database by CAD, CAPP, and CAM, respectively, when they are generating the product model. These data are further distributed to other production operations where values are added to the data to meet the requirements of the particular application. A typical example of data distribution to a specific application is shown in Figure 5.5 [3, 4]. This illustration shows how group technology (GT) is used as a means of integrating design and manufacturing by providing a common database for design and manufacturing (see also Chapter 5). This system can be implemented on its own, and at the same time, it can serve as a building block in a fully blown broad-based integrated system. In an ideal integrated CAD/CAM environment, all subsystems must communicate with one another and with other systems.

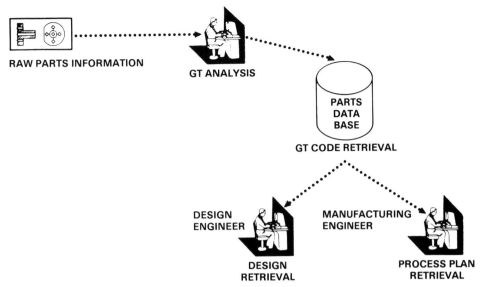

RAW PARTS INFORMATION

GT ANALYSIS

PARTS
DATA
BASE

GT CODE RETRIEVAL

DESIGN
ENGINEER

MANUFACTURING
ENGINEER

DESIGN
RETRIEVAL

PROCESS PLAN
RETRIEVAL

Figure 5.5 A Computer-Based System CAD/CAM in Operation

An Illustrated CAD/CAM System

Developments in CAD/CAM technology have made it possible to implement operational CAD/CAM systems. As previously discussed, a truly CAD/CAM system is capable of producing parts from a designer's computerized specifications. Such a system was implemented 1986 by IBM in East Fishkill, New York, having been developed to speed the design and production of prototype computer chips [4]. The system automates virtually every step in the production of a computer chip—from design to test. As an example, from several design centers around the world, circuit designers develop circuits on graphics systems. The designs are transmitted via a satellite to a quick-turn-around-time (QTAT) production facility in East Fishkill, where the designs are checked by manufacturing computers for producibility and for the production and testing of instructions.

The QTAT facility uses the generated instructions to etch the designs on silicon wafers by computer-controlled processing equipment. Other processing operations are also computerized, resulting in a system turn-around-time of one week as compared to a three-week turn-around-time using the old processing technique.

The ICAM program model served as a roadmap for integrating CAD and CAM in the QTAT facility. In this example, using the modular approach to CIMS integration, computer technology is applied to the design and production computer chips in the pursuit of a profit. The modular subsystems (CIMS) of the QTAT are tied together through a common controlled database. Data communication between the various CAD/CAM subsystems (CIMS) is accomplished through shared access to the common database.

Integrated CAD/CAM Database Systems

An integrated CAD/CAM database system is often referred to as the manufacturing database. It includes the product model data generated during the design phase and common data required by production planning, controls, processing, and shipping. Generally, much of the data used in the actual production are based on the product design (product model) data. A typical production CIMs working out of a manufacturing database is illustrated in Figure 5.6. In a manufacturing database, a designer generally creates a geometric model of the part on the screen (see Chapter 3). Additional data required to make the part are also processed and stored in the database. As illustrated in Figure 5.6, typical data stored in a manufacturing database relate to group technology, process plans, NC tool programs, inventories schedules, and the like.

A manufacturing database system consists of subsystems such as a common database, distributed databases, data communication, database management, computers, parts dictionaries, and language translators. These subsystems work together to form a reliable, efficient, and productive database system. The development of a

Figure 5.6 Production Subsystems Interacting with a Manufacturing Database

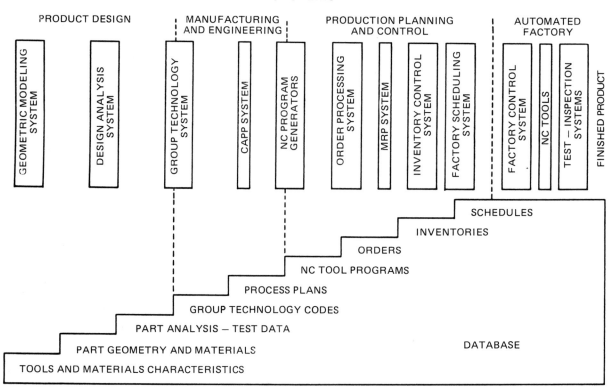

schema for an integrated CAD/CAM database is a long task. It requires a good understanding of the operations in a production cycle and a good knowledge of the future use of the database system. Because the database structure is the backbone of CAD/CAM integration, it should be well analyzed, planned, and implemented.

Common Databases

A common database contains common product data that are shared among CAD, CAPP, and CAM [5, 16]. Figure 5.7 shows the relationship among the four databases in the subsystem. The CAD database contains part design data to record geometric models, bills of materials, drawings, GT codes, documentation, and test data. Data such as group technology codes, process plans, standards, machinability libraries, tooling, and speed/feed rate tables are stored in the CAPP database. The database for CAM contains data that specify the techniques to be used in making the part. It includes such data as NC programs, orders, inventories, schedules, processing equipment, and processing. Under computer controls, the three databases interact and iterate with one another to generate the ERD, also referred to as the *common database.* This database contains the part or product model.

Distributed Databases

The common database is controlled by being signed off by engineering when the part or product model meets its desired requirements and is released to EDC. Data from the common database are controlled and distributed to other applications for their use as shown in Figure 5.8. Typical application users are manufacturing resource planning (MRPII), flexible manufacturing system (FMS), numerical control (NC), computer-aided testing (CAT), and computer-aided inspection (CAIN) and operations. Values are added to the common data to meet the require-

Figure 5.7 Common Database Subsystem Concept

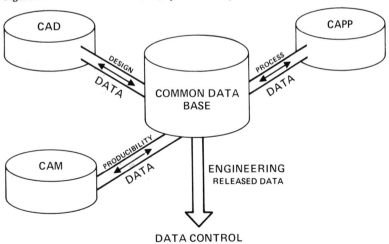

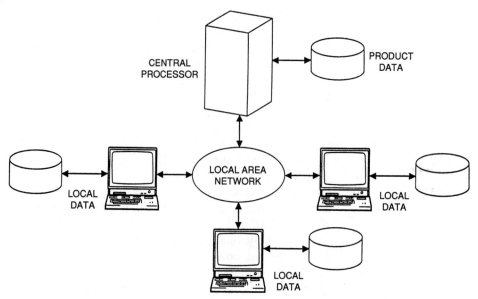

Figure 5.8 Distributed Database System

ments of the specific application. Data from the application areas are further distributed to applications at other levels for additional requirements.

A point of interest is that all data can be traced back to the common database. That is, the common database source may be viewed as a source of references or standards. The common database should be an "intelligent" database. An intelligent database will reflect any changes made in the stored data, and these changes will be reflected in the other databases in the system. For example, a change in a basic product model would be automatically carried to other applications associated with the model. Consider an engineering change that alters certain dimensions in the product model. In an intelligent database, all related dimensions on drawings associated with the model are changed automatically, as are manufacturing data, such as an NC tool path. All information is updated at the same time.

A basic key part of an integrated CAD/CAM database system is data communication. Communication links all databases together to form one integrated database system.

Database Communications

Database communications make it possible to link distributed databases and computers into an integrated CAD/CAM database system. Database communication is a subset of the CAD/CAM communication system (see Chapter 5).

A database communication network eliminates islands of automation. To the user, the distributed databases appear to be centralized. In this environment, with a good database communication network, a task may access information stored at any database(s) in the integrated CAD/CAM system. Such an integrated database permits production departments and applications to share information, which reduces redundant data in the enterprise and improves the total system's throughput. A key to solving the database communication issue is quick access to up-to-the-minute accurate data stored in the integrated database system.

The importance of easy interfaces to database communication networks cannot be overemphasized. Ports, interfaces, protocols, and controls are required for application users, engineers and designers, programmers, operators, and the like who need access to the network. Standards supported by the International Standards Organization (ISO) should be used so that equipment from various manufacturers may be attached to the network.

Database Management Systems

A database system consists of many islands of databases that are interconnected electronically and distributed throughout the manufacturing facility. Each database is managed by its own database management system (DBMS) and is generally under the control of a local host computer. A DBMS is a special software package that provides an interface between application programs and stored data, as shown in Figure 5.9. It manages the access and storage of data on a storage medium (Chapter 2). That is, a DBMS provides for access to data in ways that account for more abstract structures, such as membership in sets having certain characteristics in common.

A DBMS has many features that help users manage their data. Typical features of a DBMS are the following:

- Build physical data structures to meet the information needs of an organization.
- Control data access and prohibit access by unauthorized users.
- Simplify system restoration in the event of a failure by providing checkpoint, recovery, and restart facilities.
- Let multiple users access and/or update information in the database to make informed decisions on file placement, blocking factors, chain usage, and so on.
- Automatically log before and after update images of the database to aid in restoration if they are needed.

A DMBS plays an important role in the integrated database system. Collectively, DMBSs readily make available information on what is happening on the production floor so that the FMS will run efficiently. Types of data captured, managed, and made available to an FMS's components are process, equipment, and product status.

Computers and Database Applications

A database is a rather abstract concept. Generally, it starts out as a "bit bucket" where there is relevant information about an enterprise's products and the data needed to produce them. In an integrated manufacturing environment, however, it

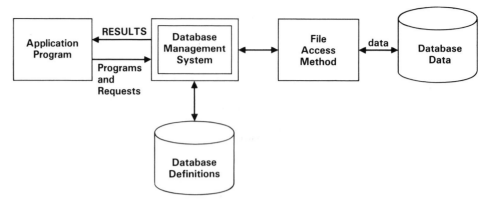

Figure 5.9 A Simple Algorithm for a Database Management System

is a complete network composed of several components. In many cases, the products are composed of various subassemblies and component parts, and each product element has both engineering and production information associated with it. To organize and manage such a collection of information in the system, a relationship between the applications, databases, computers, and languages must be understood.

New CAD/CAM applications are popping up every day to support new and established CIMS through systems integration. Typical of such applications are flexible manufacturing systems (FMS), automatic assembly operations (AAO), computer-aided testing (CAT), computer-aided inspection (CAIN), expert systems (ES), and artificial intelligence (AI) systems. Applications are also being extended in intelligent robots, automated guided vehicles (AGV), and so on. These applications use data from the shared database to meet their requirements. The key to CAD/CAM integration is to use a common, computer-integrated database system—from product development through delivery. Thus, integrity of the basic product model is preserved as values are added to this basic foundation.

Computers and Language Translators

The integration of CAD/CAM requires many different types of equipment for communicating—for example, access data and control processes. Examples include computer numerical controls (CNC) for machining centers and lathes, programmable logical controllers (PLC) for assembly and material handling conveyors, robot controls for loading and unloading the AGVs that shuttle parts between machines and between multiple cells, and an automated solidus storage/retrieval system (AS/RS) for storage of raw materials and finished goods. The integration of these subsystems requires several levels of communications, control computers, and language translators.

A translator is a special program that changes data from one form of representation to another without significantly affecting the meaning. An example of a translation process is when a program written in a high-level language is translated into the native language of the computer. Other components involved in machine-to-machine communications are lookup tables, code exchanges, dictionaries, and protocols.

A Conceptual Integrated Database System

A fully integrated CAD/CAM plant receives a "Decision to Manufacture" authorization (or WAD) to build a product. At this point, a part design is automatically coded from the information in the design database. The code in this database drives the creation of a process planning as well as other production operations. A computer-aided process planning (CAPP) system creates plans from parts codes stored in the company's GT database. Each plan specifies the sequence of production steps, tools, machining speeds and feeds, and worktime required to produce a specific part. Numerical control tool programs are generated from design data, and the fabrication and assembly for parts and tools are also prepared from design data.

Automated test and inspection systems serve a key role in the integrated system by supplying sensory feedback data that allow the system to detect and adjust automatically to changing production conditions. Robots use data from the database system to insert and solder electrical components into the boards. Inspection by means of electric test and machine-vision systems monitors production by looking for incorrectly inserted or soldered components.

To handle the entire production cycle, the integrated database system also supports production planning and control. Materials requirements planning, inventory control, and factory control databases are integrated into the database system.

Communication Systems

An integrated communication system is more than just a critical element in an integrated CAD/CAM facility. It links islands of automation, computers, databases, and various support functions such as supervision, production control, quality control, and maintenance (Figure 5.10)

The communication system is the key that unites diverse parts of a modern automated installation. Through its use in theory and in practice, the interaction of numerous discrete devices enhances each device's functionality. Communication also brings genuine automated status to the modern CAD/CAM industrial operation.

Factory communication helps to improve productivity by getting the right information to the right place at the right time. Real-time communication of events and parameter data among controllers allows automated coordination of the elements of a process. Timely communication of a product's parameters can alert plant personnel to problems and can aid in the selection of an appropriate and timely solution. Factory communication also sends to management information from various plant floor controllers and improves the accuracy of production scheduling and resource planning.

ICAM: INTEGRATED COMPUTER-AIDED MANUFACTURING

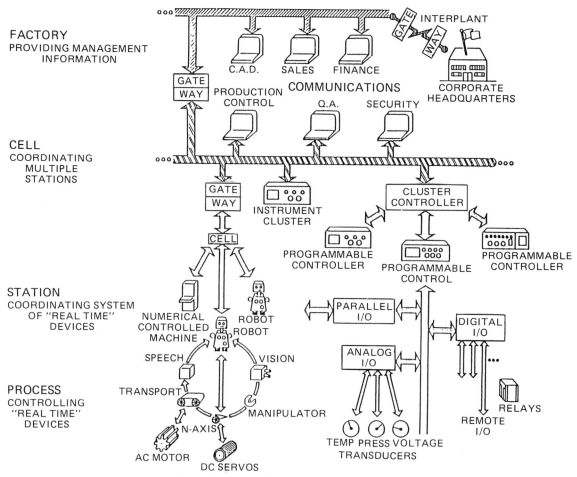

Figure 5.10 Communication in an Automated Factory

Communications must be provided at all levels of the integrated CAD/CAM factory. As a result, the needs for information and control in the facility will be different at each level. Figure 5.11 indicates the key issues at each level of the facility.

Management Communication

At the top of the hierarchy structure, Level 1, is the plant's host computer, which operates in a management environment. Communication for management information helps to support the overall operation of the business as well as the operation of the plant or process. The host computer communicates with and supports the CAE, CAPP, and CAM computers at Level 2, which are referred to as local

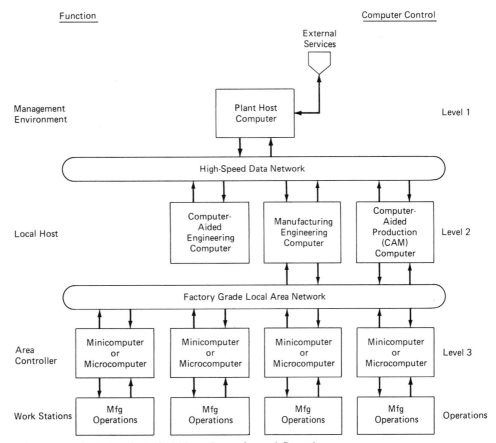

Figure 5.11 Distributed CAD/CAM Data Processing and Controls

host computers. Management information must be communicated downward from the host to the local hosts (satellite computer) and management information must be sent back from the satellite host computer up to the plant's host. A satellite host computer is used to relieve a plant host of simple but time-consuming operations such as compiling, editing, and controlling input and output devices. The satellite computer is no longer isolated from the plant host. They must work cooperatively in conjunction with computers and controllers at lower levels to achieve productivity enhancements through integrated plant-wide production controls.

A plant-level communications Local Area Network (LAN) is used to link Level 1 and Level 2 computers. This LAN is often called the backbone of the communications system. As a result, a high-speed broadband LAN is generally used. The manufacturing automation protocol (MAP), which specifies broadband communication technology (see Chapter 2) and the IEEE 802.4 token bus protocol, is rapidly gaining acceptance as the industrial standard at the plant level.

Factory Communication

Production information is communicated from local host computers (Level 2) to area controllers at Level 3 and on to workstations. Direction of production information is based on input from several functional areas. Effective communication supports the interaction of major departments and production processes. Factory communication also supports synchronization of departments, control systems, and production schedules.

A second level of LAN using MAP is at the shop floor level and links area controllers at Level 3 with local host computers at Level 2. This level of communication is referred to as a factory-grade LAN. In some cases, this LAN is a cheaper enhanced performance option of MAP known as MAP/proway. It uses carrier-band technology instead of broadband, and it streamlines the use of the ISO model by removing the middle layers.

Planning and Implementation

Planning

An integrated communications system must be carefully planned at all levels with the company's goals in mind. Successful automation is a gradual process, and a plant's evolution toward integrated communication can be likened to the building of a pyramid (Figure 5.12) [6]. Automation starts when controls are interfaced to the machine and process equipment at the pyramid's base. Implementation occurs where the control systems meet the machine and process equipment. Increased production rates place greater demand for response times in the milliseconds.

Machinery and processes respond to commands from the station-level controls. Therefore, access to plant-flow data is necessary for supervisory monitoring and control, and it must be formatted for easy understanding by plant personnel. It must be available in real time to depict plant floor conditions accurately. As stations multiply, the call level comes into play, coordinating their functions to allow integrated monitoring and control. Control messages should have priority over information messages to ensure the safety and reliability of the control system yet maintain the required access to real-time data to support decisions.

Devices on the center level coordinate multiple cells for scheduling, production, and management information. At the top of the pyramid shown in Figure 5.12, the plant level, management directs planning, execution, and control of plant operations. A key factory communication issue is information management within and among manufacturing and process areas. Typical questions that may be asked for this level are: What is the right information for plant floor coordination? Where should it be and when? How should it be acted on?

Implementation

Implementing an integrated communication system that addresses the planning issues we have discussed is not an easy task. An approach to an implementation solution places the issues into six functional areas [7, 3]:

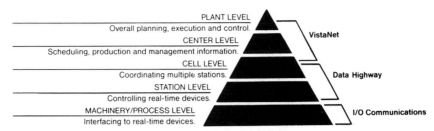

Figure 5.12 A Productivity Pyramid

1. Distributed control: for integrating devices on the plant-floor level within rigid response times while still enabling upward communication to coordinate the entire process.

2. Monitoring: for gathering uptime and downtime information, alarm histories, trending parameters, and other data in a nonreal-time environment.

3. Data acquisition: for using the network to provide access to an on-line database related to the manufacture of the specific product. (Parts counts, parts rejected, and other quality and production data are included in this database.)

4. Supervisory control: for supplying *actionable* information to higher levels and returning the appropriate response. (An alarm may notify an operator that something is out of tolerance, and it may request a specific command. The operator response completes the supervisory control process.)

5. Program support: for using the communication system to upload, download, and store programs (transferring very large files, creating high-capacity but less time-critical demands on the communication system).

6. Management information: for gathering large amounts of preprocessed information from lower levels to facilitate batch transfer to the plant computer.

A Communications Model

A factory communication system should permit all the facility's industrial devices such as PLC, NC, weld controllers, robots, vision systems, and the like to communicate with one another and the computer equipment. In most cases, the facility is divided into areas of functionality. As a result, no single network can serve all these areas; each area usually has its own computer system. In many cases, each area is equipped with devices supplied by various manufacturers and installed at various times. Two questions that are generally asked are (1) How can these devices be made to communicate with each other? (2) How can we avoid a failure to communicate?

Many manufacturers are designing systems to address these two questions. Allen-Bradley addresses the questions by using the scheme shown in Figure 5.13 [7]. VistaNet is a broadband network that distributes production data (communicated by controls over data highway and data highway II) both vertically to the upper-level computers and horizontally to area control computers. The functional units shown in the figure are the following:

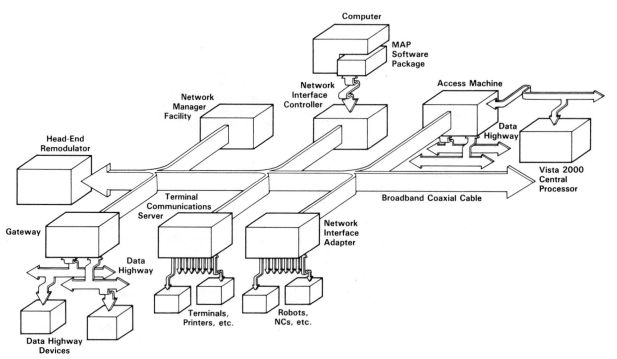

Figure 5.13 A Communications Network (VistaNet LAN Configuration)

1. Head-end remodulator. The head-end remodulator acts as a relay between the receive and transmit channels of the VistaNet LAN. The unit is needed to achieve full 802.4 compatibility within the VistaNet LAN.

2. Network manager facility. The network manager facility allows one to perform a variety of functions on the VistaNet LAN from a central location and is required for the Allen-Bradley VistaNet LAN.
   • Monitor the VistaNet LAN.
   • Collect and analyze performance statistics.
   • Perform diagnostics.
   • Download VistaNet node software for maintenance and MAP upgrades.
   • Generate, maintain, and download configuration files.
   • Read and set communications layer parameters.

3. Gateway. The gateway connects two data highways to the VistaNet LAN. The data highway is a baseband local area network specifically designed for industrial applications.

   The gateway allows stations on the two data highways to communicate with each other and with others on the VistaNet LAN. The gateway also allows a method by which a system can be structured with the distinct network characteristics required to fit specific needs, and it facilitates the integration of multivendor equipment into a local area network system.

4. Network interface controller (NIC). The network interface controller connects a

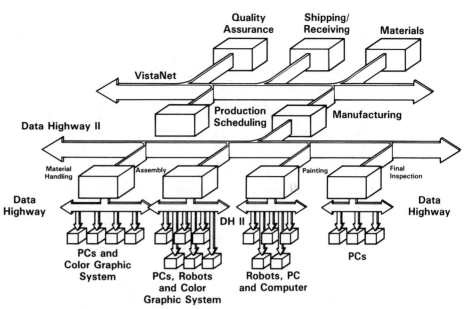

Figure 5.14 VistaNet Interfaces

variety of computers and other intelligent devices to the VistaNet LAN. The NIC provides a full duplex, high-speed, general-purpose interface to the VistaNet LAN.

5. Network interface adapter. The network interface adapter connects eight factory automation devices, such as robots and weld controllers, to the VistaNet LAN.

6. Terminal communications server. The terminal communications server connects 12 RS–232/RS–422 peripheral ASCII devices (e.g., terminals and printers) to the VistaNet LAN.

7. Access machine. The access machine interfaces the Allen-Bradley Vista 2000 central processor to the data highway and coordinates the flow of information between the data highway, the data access link, and the VistaNet LAN.

8. The MAP software package. The MAP software package provides layers 3 (network) through 7 (application) of the VistaNet LAN architecture. A complete VistaNet LAN interface is created for computers or intelligent devices when MAP software is used with the NIC.

VistaNet

VistaNet is the interarea connection. It integrates separate automated production systems such as robots, color graphics, PLCs, NCs, and the like, but the goals are quality assurance, shipping/recieving, production scheduling, and material management, not production itself (Figure 5.14) [7]. The network sends the production data transmitted over the data highway and data highway II to the upper levels of the plant computer hierarchy.

VistaNet connects devices that interpret collected data, communicates this information to appropriate supporting functions, makes decisions to direct process, and feeds back the decisions to optimize production. Special devices may be configured with VistaNet to form an Allen-Bradley Vista Area Management System. VistaNet is designed to interact with the products of multivendors. It is a family of products that interconnect factory floor controllers, data entry stations, printers, terminals, and computers. It transmits data at 5 million bits per second with a high degree of noise immunity.

Data Highway and Data Highway II

A data highway (Figure 5.15) is an industrial area network for plant control applications that is capable of connecting as many as 64 intelligent devices [7]. These devices may be connected in various combinations to communicate as peers. As a result, the designer is not limited to connecting only one computer to the network or forced into choosing between connecting a computer or a color graphic system to the network. A data highway is easily expandable, giving the designer the flexibility of not being forced to connect to all these devices at once. Stations may be connected to the network by drop lines extending up to 100 feet. The stations may be as close together or as far apart on the main trunk line as desired.

Industry-wide data highway applications include the following:

- Bar code data transmission to verify parts assembly.
- Ingredient disk pack, based on recipes downloaded from computers to programmable controllers.

Figure 5.15 Data Highway Interfaces

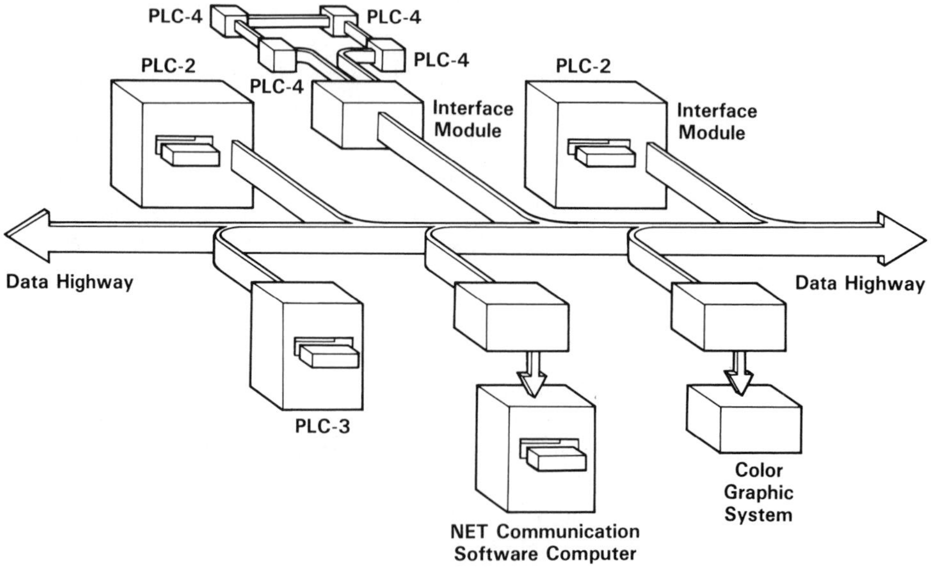

Fundamentals of Product Processes and Operations

- Paper production trucking and diagnostics.
- Keeping track of material routing information and set points for batching and weighing systems.

Data highway II (Figure 5.16) is a flexible, high-speed time system of fast response that is optimized for all network applications [7]. It addresses the need for high performance and easy interaction among different types of programmable devices. Also, data highway II satisfies higher-level needs in the productivity pyramid shown in Figure 5.12. It provides for message priority and high throughput at each node to transfer information in and out quickly. The data highway II provides a solution for applications requiring higher performance than the data highway to address communication issues at the station and all levels.

In summary, this communications model serves as a guide in the design of communication networks in a factory control system. It shows how various communication solutions concerning the issues can contribute to the success of the strategies planned for a production facility's future.

Process Planning

Process planning is that phase of manufacturing in the production cycle (Figure 1.7) in which product designs are translated into the processes required to produce the product. This function is the prime interface between product design and production (Figure 1.9). As a result, process planning is responsible for the general flow of information from design engineering to the factory floor.

Figure 5.16 Data Highway II Interfaces

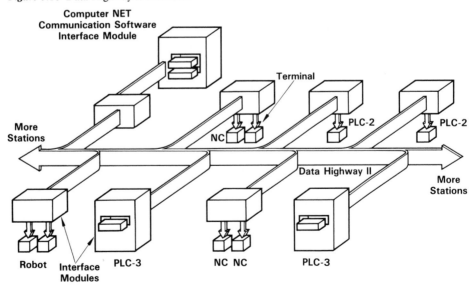

A process plan typically contains such items as the serial steps of operations to be performed, the machines or processes to be used, the cost centers involved, labor standards for setup and operation run time, and any tools, gauges, or information required. In some cases, manufacturing professionals refer to the "process plan" as the "production design."

Items in a process plan come in various forms. The route sheet is expressed in one form. A route sheet contains a sequence list of the individual production operations required to make a part. Also included in a route sheet is a list of the associated machine tools and equipment needed to perform the operations. Another item is represented by NC processes and programs. It is prepared by a parts programmer for that portion of the jobs to be accomplished by NC machines.

Traditionally, process planning tasks are carried out with a high degree of manual and clerical work. Many of these tasks and routines in one planning operation are similar or identical to those in other operations. As a result, many tasks and routines are repeated over and over. Problems encountered with manual process planning are enormous and have resulted in the automation of many activities in this function. Automated process planning activities are a high level of CAPP.

Computer-Aided Process Planning (CAPP)

The use of computer resources to aid the process planner in a systematic determination of proper methods to be used in the production processes is called CAPP. These systems manage the storage, retrieval, distribution, and maintenance of the process plan library. They have been extended to include the logic used by process planners in choosing alternate process methods. A key to the development of CAPP is to structure the data concerning parts fabrication, facilities, tooling, and materials into categories and logical relationships. The computer can then make the many comparisons necessary to create the optimum plan.

A CAPP system links the information transformation process between CAD and the requirements of MRP II (Chapter 6) and CIPM. Both MRP II and factory automation areas require a "road map" detailing the routing and timing of the manufacturing process.

Modern CAPP systems are focused around two technologies: Variant and Generative. The major difference between the two is the manner in which the manufacturing knowledge is stored.

Variant Systems

A variant system uses a GT code for the new part to search the database and to identify like or nearly alike process plans. The GT code identifies and brings together related or similar components and processes to take advantage of their similarities in design and production. Variant CAPP systems also support the use of standard plans that are created for a family of parts and then customized for a new part. The old plans become the reference base for making planning decisions.

A variant CAPP system is basically an editing system that enables a process planner to create a new plan by retrieving and modifying an existing, manually prepared plan. That is, in a variant CAPP system, a process plan is created through the modification of either a generalized standard plan for a specific part family or an existing plan for a similar part within a specific family.

Generative Systems

A generative system is similar to an expert system (see Chapter 5). An expert system is a computer program that draws on the organized expertise of one or more human experts. The rules for creating a process plan are codified, and the planner is led through the steps to generate a new process plan. The logic embodied in the old plan is formalized into decision trees. As a result, instead of modifying existing plans, a generative system creates a totally new plan.

To create a totally new plan, the rules and logic of the manufacturing processes must be captured and stored in a database. That is, a generative approach uses the computer to synthesize each individual plan. This technique uses the appropriate algorithms that define the various technological decisions that must be made. It relies on the computer's memory, logic, and computational powers. The planner in such a system performs merely a monitoring function and arbitrates some conflicts about elementary decisions.

Even though most of the CAPP systems sold today are variant, current market trends are beginning to focus on generative systems. There are some major advantages of taking the generative approach: consistent plans, planner independence, the use and retention of the optimum manufacturing logic, compatibility with new technology, overall quality, and vastly improved machine use. Also, because this system does not rely on stored plans, it has unlimited flexibility and greatly diminished requirements for data storage.

CAPP Systems Operations

Lockheed's CAPP System

Lockheed's CAPP system, called GENPLAN, is unique in its ability to process plans from the beginning [8]. This system creates plans from parts codes stored in a company's GT database. Each plan specifies the sequence of production steps, tools, machining speeds and feeds, and worktime required to produce a specific part. A plan for a simple part, such as an aluminum mounting bracket, takes only minutes to generate as compared to several hours it would take a process planner to do the same job. Thus, this system improves productivity through reduced lead time for planning, improved accuracy, and standardization. It also provides interfaces to accept part design from CAD systems and pass information to CAM systems. This operation is illustrated in Figure 5.17.

In providing the necessary links, CAPP accepts the product definition information from CAD systems or an engineering database, and determines how the product is to be produced. It then passes the processing information to the appropriate CAM systems (CAP) or to the production database. The actual information accepted, processed, and passed to CAM systems is dependent on the type of product designs.

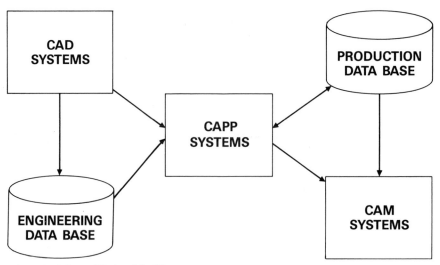

Figure 5.17 Central Role of CAPP

*Organization for Industrial Research, Inc.'s
MULTICAPP System*

The Organization for Industrial Research, Inc.'s (OIR) MULTICAPP is a modular system with a great deal of user-oriented flexibility [3, 17]. It provides CAPP on several levels in a manufacturing environment. As an example, MULTI-CAPP produces finished process plans for use on the shop floor with a minimum of clerical work by the process planner. This system works automatically, retrieving existing process plans (based on company experience) for same or similar parts, and allowing the process planner to make changes, if needed, to conform to a specific part. A typical MULTICAPP process plan is shown in Figure 5.18 [3, 18]. Such a process plan is created from scratch, using standard process descriptions contained in the system text file. The text file is then assembled and edited for each step in the process. All this is done interactively through the computer. In effect, the MULTI-CAPP system serves as an "electronic pencil" used to edit instructions and create new ones.

OIR's MULTICAPP also provides for pictorial process planning as shown in Figure 5.19 [3, 21]. Pictorial process planning integrates CAPP and graphics. This system makes it possible to create detailed process plans that include both text and illustrations for every operation (Figure 5.19). The output includes detailed routings and step-by-step drawings to accompany routing instructions.

Summary of Computer-Aided Process Planning

A CAPP system may be viewed as the use of computer resources to improve productivity of the process planner. It is an extensive system designed to improve productivity in the interface between design and production and manufacturing engineering. It was developed to communicate the production process to be used to

Fundamentals of Product Processes and Operations

Figure 5.18 A MULTICAPP Process Plan (OIR)

ORGANIZATION FOR INDUSTRIAL RESEARCH, INC. FACILITY – F1								

PART NUMBER: PROB.15.10.1	LAST FOUR ORDERS			MINIMUM QTY	DUE DATES	PRI #
	S/O #	PRJ #	QTY			
PART NAME: DRIVER,VLV GUIDE						
PLNG REC: DWG REV: C	4-232	SD122	4000	3500	12-81	1
PLANNER: FRED SAMBERA						

CHANGE APPROVALS & DATE				CODE #1: 1-3300-07-234901-5-0516-0000000000000
	#1	#2	#3	CODE #2: 5-2120-3654-22-01
MG ENG Q/A		A3 E1 Q2	E2	CODE #3: 6-4032-417
				START: 08/15/81 IT.Q.T.D.: 4000 IT.R.DOC: 1

MATERIAL REQUIRED:
SPECIAL INSTRUCTIONS:

OPER NO	MACH TOOL	OPERATION DESCRIPTION – ASSY INSTRUCTIONS	TIMES		OPERATOR STAMP
			S/U	RUN	
0010	1258	SET-UP 3/4 DIA COLLET PADS SET TURRET STOP TO HOLD 4.5 LENGTH ROUGH TURN .5 DIA TO .532 DIA +.01 −.01 ROUGH TURN .375 DIA TO .39 + .01 −.01 HOLD 1.625 LENGTH FINISH TURN .500 DIA + OR −.005 FINISH TURN .375 DIA + OR −.005 HOLD 1.625 CUT-OFF TO 5-7/8 LENGTH	1.7		.40
0020	1258	S/U COLLET HOLD ON .500 DIA ROUGH TURN .437 DIA AND FORM 30 DEG ANG. FINISH TURN .437 DIA + OR −.005 AND FORM 30 DEG ANG HOLD 1.75 DIM + OR −.015 FACE TO 5 3/4 LENGTH + OR −1/16 BURR SHARP CORNERS	1.5		.15
0030	9401	HARDEN, HEAT AT 1550 DEG. F. − OIL QUENCH AT 120 DEG. F.	.25		.50
0040	9401	STRAIGHTEN TO .005 T.I.R.	.10		.25
0050	9401	TEMPER AT 400 DEG. F. MIN. TC 46-50 R.C. FOR ONE HOUR	.25		0.05
0060	9805	INSPECT HEAT TREAT − 46-50 R.C.	.10		.15
0070	9201	CHROME PLATE	.27		.05
0080	4102	S/U THRU FEED GRIND 00 TO .5100	.375		.06
0090	9805	INSPECT ALL DIMENSIONS PER PRINT	.10		.15

Integration of CAD and CAM Technologies

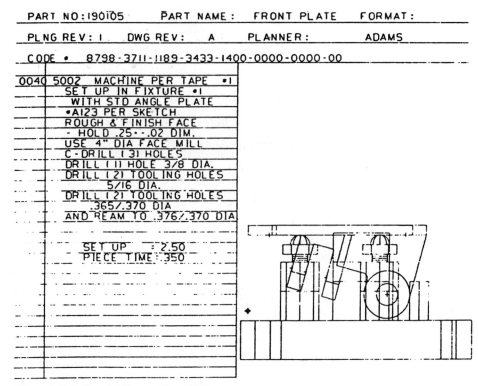

PART NO: 190105   PART NAME: FRONT PLATE   FORMAT:

PLNG REV: 1   DWG REV: A   PLANNER: ADAMS

CODE • 8798-3711-1189-3433-1400-0000-0000-00

0040 5002 MACHINE PER TAPE •1
SET UP IN FIXTURE •1
WITH STD ANGLE PLATE
•A123 PER SKETCH
ROUGH & FINISH FACE
- HOLD .25•-.02 DIM.
USE 4" DIA FACE MILL
C-DRILL (3) HOLES
DRILL (1) HOLE 3/8 DIA.
DRILL (2) TOOLING HOLES
5/16 DIA.
DRILL (2) TOOLING HOLES
.365/.370 DIA
AND REAM TO .376/.370 DIA

SET UP = 2.50
PIECE TIME = .350

Figure 5.19 A Pictorial Process Plan (OIR)

produce a part to the factory floor. As manufacturing technology moves toward more automation, it is mandatory to optimize operations and processes at the plant level. To do this, CAPP should be integrated into factory management to meet the enterprise's goals at the higher level. From a production planning point of view, CAPP systems help with analysis to facilitate cutter life forecasting, materials requirements planning, scheduling, and inventory control.

The use of CAPP simplifies a time-consuming, knowledge-intensive task and results in greater consistency. A CAPP system positively influences machine use, capacity planning, production scheduling and lead times, cost estimating, and producibility. The interaction of CAPP with certain functions in the production cycle is shown in Figure 5.20. This environment permits CAPP to serve as a bridge between product design and production, making CAPP an indispensible element in an integrated CAD/CAM environment.

Group Technology (GT)

Group technology is a philosophy based on a fundamental principle of identifying and bringing together related or similar attributes to achieve efficiencies by grouping like problems. In most cases, a prerequisite for the recognition of similarities is a

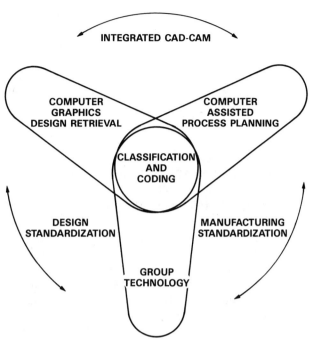

Figure 5.20 CAPP Working Environment in the Production Cycle

system by which the objects can be classified and coded. The GT concept is currently attracting a lot of attention from the manufacturing community. In manufacturing, GT identifies and brings together related or similar attributes of parts, processes, and other pertinent data entities. This organization principle is used to take the advantage of similarities in attribute in such functions as design and production.

Typical attribute groupings in manufacturing are production cells, geometric properties, component materials, production processes, methods, and so on. Attributes groupings between objects of interest are very popular for the Big Three of the production cycle.

Product Design

In product design, parts are classified and GT coded (see Chapter 5). According to geometric similarities by using codes that have certain attributes. Such a grouping of part forms is what is known as a "family of parts." Engineers classify parts and assign closely related attributes to a particular family of parts so that they can determine similarities among them in several ways. For example, similarity can mean closely related geometric shapes and dimensions. The designer can retrieve all parts having certain features, such as rotational parts with a length-to-diameter ratio of less than 2. If one of these fits the need at hand, the designer can thereby avoid having to design a new part.

Process Planning

In a like manner, GT coded objectives can be used by process planning. The GT code stored in a GT database speeds up the retrieval of parts information, facilitates the process planning, improves the accuracy of process planning, aids in the creation and operation of manufacturing cells, and enhances the communication between functional areas.

Production

Similarities between parts that are captured in the GT code can be used in production. Historically, GT has been associated with manufacturing cells in which machines used in the production of families of similar parts are grouped together to accomplish a more efficient flow of material and to reduce set-up times. Also, production can drastically reduce the time and effort spent deciding how a part should be produced if this information is available for a similar part. Similarity between two parts means that they are processed through the factory in the same, or almost the same, way. This does not mean, however, that if parts look alike they are always produced in the same way. That is, parts routed through the same machines can be quite dissimilar in geometric form.

Classification and Coding

Classification may be defined as gathering like items such as parts, materials, processes, tools, and so on by their similarities and separating them by their differences. That is, similarities and differences determine the logical breakdown (classification) of the item population.

A classification system should be viewed objectively with a clear understanding of all the planned applications of the classification system and its related database. Figure 5.21 shows a logical structure of parts based on their shape and other physical attributes (2). This logical structure is suited for a coding scheme as discussed in Chapter 5 and is referred to as a logic tree (rotate figure 90 degrees clockwise to the right). Logic trees can serve as pointers to files or families of existing designs stored in a database.

A code may be defined as a set of rules describing unique notations arranged in a clear way so that it can be translated by a system. An example of a coding scheme is illustrated in Figure 5.22. This code can easily be applied to the logic structure in Figure 5.21 where D1, D2, D3, D4, and D5 represent the various physical attributes of the part. Several proprietary systems with software are currently being marketed. Consider the implications of combining applied group technology to the design and production processes. Parts with similar characteristics, both from a design and production point of view, are grouped together in families of parts. The result is a significant reduction in the number of unique work problems with which design and production functions must deal.

To take advantage of the principles of GT, there must be a means of accomplishing the grouping of parts into families. This is the classification procedure that

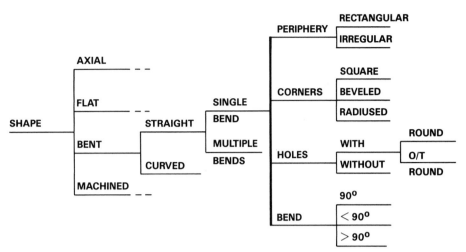

Figure 5.21 Structure of Parts Based on Manufacturing Process and Shape (A Tree Diagram)

identifies the design and production characteristics of each part and thereby serves as the mechanism for forming these families.

Coding Schemes

Numerous coding systems have been developed all over the world by university researchers, consulting firms, and corporations. The systems range from simple informal techniques to highly structured formal techniques. An informal parts classification technique was developed with the sole intent to identify families with similar production requirements to create dedicated lines or cells of machines.

Even though informal ways of grouping parts are not uncommon, the greatest potential of GT in manufacturing comes from a formal coding system. A formal coding system assigns each part a numeric or alphanumeric code describing the attributes of interest.

Parts classification and coding systems are divided into one of the following four general categories:

1. CAD—Systems based on part design attributes
2. CAM—Systems based on part production attributes
3. CAB—Systems based on part business attributes
4. Combinations:
   - CAD/CAM—Systems based on both design and production attributes
   - CIM—Systems based on business, design, and production attributes

Even though many areas of business operations can benefit from GT, operations in the production cycle are where it is most highly practiced. Three major operations in the production cycle where GT is highly used are process planning,

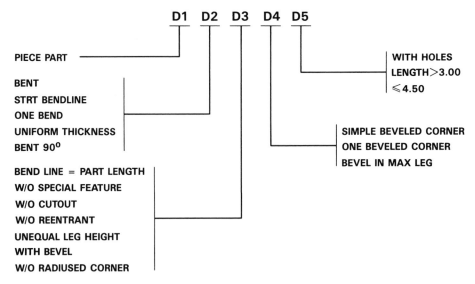

Figure 5.22 A Five-Digit Part Family Code

production planning and control, and design. In CAD/CAM, the combination category is useful. In this category, the functions and advantages of CAD and CAM systems are combined into a single classification scheme.

A coding scheme for parts consists of a sequence of numeric and/or alphanumeric characters that identify the part attributes. Such a classification scheme serves as a mechanism for forming part families.

Coding Structures

Three basic types of GT code structures are commonly used:

1. Hierarchical structure
2. Chain-type structure
3. Hybrid structure

Hierarchical Structure

The logical structures shown in Figure 5.21 are hierarchical. In a hierarchical structure, the interpretation of each succeeding symbol depends on the value of the preceding symbol, as illustrated in Figure 5.23. The attributes of a workpart are coded as illustrated in Figure 5.24. Assigned physical attributes of the workpart are shown by the heavy line on the tree (Figure 5.24B). The translated code is shown in Figure 5.24C.

In many cases, the hierarchical code structure is used interchangeably with *monocode* and *tree structure codes.* Hierarchical code structure is the oldest type of

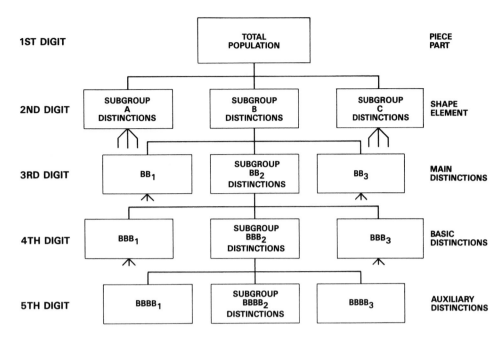

1ST DIGIT — TOTAL POPULATION — PIECE PART

2ND DIGIT — SUBGROUP A DISTINCTIONS / SUBGROUP B DISTINCTIONS / SUBGROUP C DISTINCTIONS — SHAPE ELEMENT

3RD DIGIT — BB₁ / SUBGROUP BB₂ DISTINCTIONS / BB₃ — MAIN DISTINCTIONS

4TH DIGIT — BBB₁ / SUBGROUP BBB₂ DISTINCTIONS / BBB₃ — BASIC DISTINCTIONS

5TH DIGIT — BBBB₁ / SUBGROUP BBBB₂ DISTINCTIONS / BBBB₃ — AUXILIARY DISTINCTIONS

Figure 5.23 Hierarchical Code Structure

Figure 5.24 A Coded Workpart

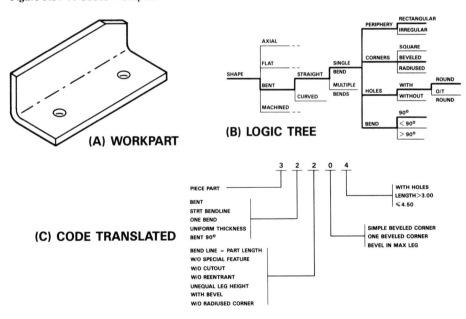

(A) WORKPART

(B) LOGIC TREE

(C) CODE TRANSLATED

3 2 2 0 4

PIECE PART

BENT
STRT BENDLINE
ONE BEND
UNIFORM THICKNESS
BENT 90°

BEND LINE = PART LENGTH
W/O SPECIAL FEATURE
W/O CUTOUT
W/O REENTRANT
UNEQUAL LEG HEIGHT
WITH BEVEL
W/O RADIUSED CORNER

SIMPLE BEVELED CORNER
ONE BEVELED CORNER
BEVEL IN MAX LEG

WITH HOLES
LENGTH>3.00
≤4.50

coding. It provides a relatively compact structure that conveys much information about the part by using only a few digits. This makes it very effective when used for retrieval purposes in a large database.

Chain Structures

In the chain type of structure, in contrast with the hierarchical structure, individual symbols in the number sequence are fixed and do not depend on the value of the preceding digits. This code is usually long. Every digit in the code represents a distinct bit of information about the total part. For example, one digit may be used to define form, the next materials, the next dimension, and so on. By chaining these building blocks together, a complete part can be described in a manner that is internally consistent with the description of all parts in the system. An example of how this code is structured is illustrated in Figure 5.25 [9]. All part features are identified by a specific code character, which leads to some extremely lengthy codes. A typical code of a workpart is also illustrated in Figure 5.25.

The chain type of code is often referred to by other names, such as polycode, attribute code, fixed-digit codes, and descriptor code. Regardless of the name, this code lends itself nicely to the formation of part families and the implementation of GT techniques. Polycodes are well suited when parts are to be classified by technological processes.

Hybrid Structure

A hybrid structured code is a combination of the monocodes and polycodes, as illustrated in Figure 5.26. This code is often referred to as a multicode code and is popular in many GT systems.

An advantage of the hybrid code is that it is generally constructed to achieve the best features of the two pure codes. As an example, hybrid codes are typically constructed as a series of short polycodes (Figure 5.26). The digits in the series of short codes are independent, whereas one or more symbols in the code set are used to classify the part population into groups as in the hierarchical structure.

Implementation of Group Technology (GT)

Group technology calls for simplicity and standardization. In CAD/CAM, it is a technique for identifying and bringing together related or similar components to take advantage of their similarities in the design and production process. It has been called the glue that holds CAD and CAM together. It is a concept that is attracting a lot of attention from the manufacturing community. The essence of GT is to capitalize on similarities in recurring tasks by doing the following:

- Performing similar activities together, thereby avoiding wasteful time in changing from one unrelated activity to the next.
- Standardizing closely related activities, thereby focusing on distinct differences only and avoiding unnecessary duplication of effort.

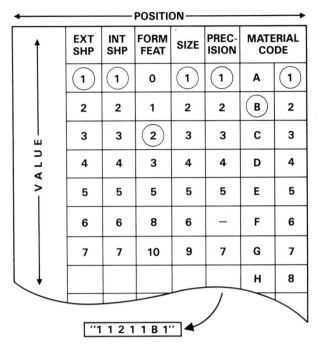

Figure 5.25 The Chain Type of Code Illustration

Figure 5.26 A Hybrid Structure Code Format

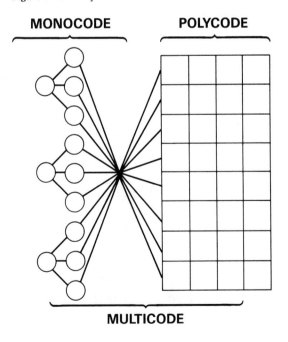

- Efficiently storing and retrieving information related to recurring problems, thereby reducing the search time for the information and eliminating the need to solve the problem again.

Group technology requires careful and detailed planning before implementation. Planning starts with the objectives of the enterprise and the applications of GT in the manufacturing process. The GT codes are selected to meet the requirements of the designated functions. Management plays an important role in the formulation of strategic plans and must be more familiar with the applications and benefits of GT than at present.

Applications of GT

Typical application areas of GT are scheduling, process planning, creation of families of parts, sequencing of parts families on machines, parts designs and modifications, creation of production cells, purchasing, cost estimation, and determination of the economic consequences of anticipated changes in materials cost.

Benefits of GT

Group technology is a means of integrating manufacturing applications. Its prime role is to integrate design and production activities, as shown in Figure 5.27 [10]. Applications of GT offer the potential of improving workflow, throughput, work in process, stocks, and machine use through such benefits as the following:

- Retrieval of existing valuable engineering documentation and analysis on parts.
- Avoidance of duplication of part design.

Figure 5.27 CAD/CAM Integration Through GT

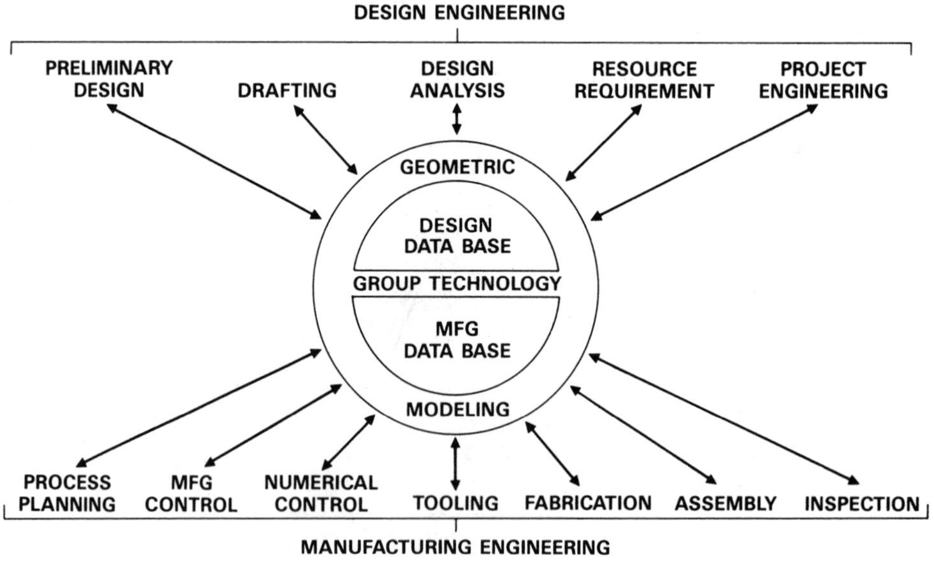

Fundamentals of Product Processes and Operations

- Reference to existing design on similar parts.
- Development of engineering standards.
- Elimination of process variance in manufacturing engineering.
- Standardization of the most efficient manufacturing processes.
- Interdivisional cost and manufacturing capability comparisons.

Expert Systems (ES)

An ES is a computer program that draws on the organized expertise of one or more human experts. The computer takes the expert advice of humans, coded as a series of rules, and applies it to a specially structured base containing information about a real or hypothetical situation. In these systems, a large volume of knowledge is entered into a computer system from human experts. By taking actions specified by the rules, the computer simulates the behavior of human experts in confronting the same situation.

Expert systems are a common application of artificial intelligence (AI) in such industries as electronics, medicine, chemistry, and geology. Artificial intelligence is that branch of computer science dealing with symbolic, monalgorithmic methods of dealing with problems.

Artificial Intelligence (AI)

Artificial intelligence is a complex subject. As a result, it is difficult to define. In addition to the definition given in the preceding section, a variety of different and somewhat controversial definitions can be found in the literature. Typical of them are the following:

AI is the development of computer systems capable of mimicking human reasoning and perception.

AI is the process of symbolic information such as concepts, knowledge, and relations.

AI is the study of techniques for solving exponentially hard problems in polynomial time.

AI is the part of computer science concerned with designing intelligent computer systems. [21]

For the purpose of discussing ES, AI is defined as a study of computer science concerned with the modeling of human faculties to do things at which, at the moment, people are better. By understanding AI from this point of view, you will gain knowledge of ES in terms of applications, methods, and techniques of constructing man-made intelligent machines. Intelligent machines are the focal point and are of interest in the factory of the future.

A typical use of AI is in automation systems driven by manufacturing production data. Data automation systems are used in such functions as CAPP, CAPM, CAP, and training. Such functions require extensive thinking, time-consuming, and labor-intensive processes. With labor costs in such functions approaching a high percentage of the total production costs, a major emphasis of AI in expert systems is focused on such activities as planning, methods, processes, standards, cost estimating, and scheduling. Experts systems can augment, if not automate, these activities.

Experts Systems in CAPP

An ES is used in process planning because this function encompasses a broad spectrum of manufacturing support activities such as planning, methods, processes, and standards as well as other aspects of manufacturing activities such as cost estimating, scheduling, producibility, inspection, and tests. These activities are natural extensions of CAPP.

A use of expert systems in CAPP activities form knowledge-based production systems that add a high degree of intelligence to the systems. The degree of intelligence of the system depends on the intelligence level of the expert system. A knowledge-based system is ideally suited as an automated process planning system.

An automated process planning system is the purest form of generative planning. With this system, manufacturing process plans are derived directly from the engineering database. Unlike CAPP, which emphasizes the augmentation of manual procedures, automated process planning emphasizes the elimination of manual procedures.

Knowledge-based automated systems facilitate the implementation of generative process planning by alleviating the time-consuming and labor-intensive chore of modeling each and every unique situation. This is accomplished by teaching the system how to deal in generalities and by providing the system with certain basic rules that deal with how to recognize patterns and make assumptions. Such knowledge-based systems are powerful systems with a high degree of automation capabilities. However, these systems have limited flexibilities in problem solving as compared to human beings.

Flexible Manufacturing Systems (FMS)

Overview

Manufacturing Systems

A basic framework for an FMS is a well-defined manufacturing system. A manufacturing system usually consists of a complete set of processes, resources, and equipment brought together for a specific purpose. Such a system is used to create a specific set of discrete products. The created products may vary in size from integrated circuits (IC) smaller than a fingernail to an entire automobile. They may be as simple as a fiber washer or as complex as an NC machine. The system may be

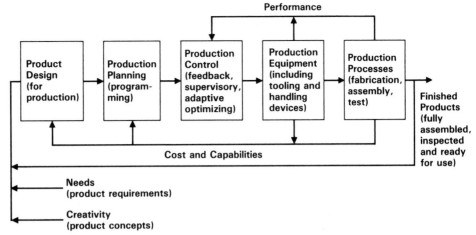

Figure 5.28 CAD/CAM Paves the Way for the Automated Factory

comprised of a single work cell dedicated to specific processes or several work cells each of which may be different processes and operations.

Typical production and control systems (such as CAPACS and CIMs) integrated into a production environment to form a manufacturing system are shown in Figure 5.28 [11, 8]. This figure focuses on industrial applications of discrete manufacturing. It exemplifies a manufacturing system comprising several work centers, each dedicated to different production processes and operations. That is, this system is composed of a multiple number of distinct processes and operations that are being performed on a product—from design to completion, as discussed in Chapter 5.

Flexible Manufacturing

One of the most exciting concepts for the automated factory is FMS, which commonly refers to computer integration of the many individual automation concepts and technologies into a single productive system. A characteristic of this productive system is the ability to produce simultaneously different kinds of parts. This system also describes a wide variety of systems containing a broad range of flexibility and automation. As a result, FMS can mean any automation application from the use of CNC machines or robots, to entirely automated factories, as illustrated in Figure 5.3.

The FMS symbolizes a new area in manufacturing—an area of flexibility. This flexibility must be realized through reductions in the numbers of machines, the amount of contingency stocks, the process set-up losses, and other costs caused by a lack of flexibility. The key aspect of FMS is its ability to adapt to change and not just the degree of automation it incorporates. It is flexible in terms of the quantities of parts it can handle, the types of products it can produce, the order in which processing steps may be performed, and its ability to reroute parts back into flow paths. It also fills the gap between high-production transfer lines and low-production machines such as NC.

Flexible manufacturing requires real-time information and decisionmaking to take advantage of the built-in flexibility of the system. As an example, production operations in larger factories usually have multiple supervisions, multiple products, and hundreds of parts in the materials inventory. The flow of information between operations requires a real-time exchange of data to maintain an effective flexible system. Flexible manufacturing goes beyond the walls of functions and factories to provide this real-time exchange of data for FMS.

A truly integrated CAD/CAM factory provides for the exchange of information during its operation. It is important that CAD and CAM be integrated to take advantage of the FMS. Integration of CAD and CAM technologies into an overall FMS can result in a major increase in productivity, reliability, and repeatability through the reduction of manual intervention and precise process control. The system can also reduce total disk throughput time, thereby decreasing work-in-process inventory. Disk throughput is the time it takes to retrieve data from the disk system after it has been called for by the program.

Elements of FMS

The thrust of FMS is focused on the theory of GT and its applications in three main categories.

- Machine tools
- Materials handling systems
- Computer control system

Although consideration must be given to the selection of machine tools and the materials handling system, these are mechanical systems that depend on the processing requirements and that perform their tasks if the plans and support systems are in place. Thus, the computer control system is the linkage needed to transform a group of machines and standalone systems into an effective FMS.

The functions performed by the computer control system may be categorized into either machine support or planning support. Typical machine support functions are NC programs for the storage of parts, for the distribution of parts to the individual machine tools, for traffic control, and for tool control. These functions are required to support the machine directly. Functions such as routing, scheduling, and the monitoring and reporting of system performance are needed to plan the employment of resources.

Expert Systems in FMS

Knowledge-based systems are being increasingly used within automated manufacturing in production processes, operations, and maintenance. These systems provide opportunities for productivity improvement by making valuable knowledge available to a wide range of users who may have little access to the cumulative

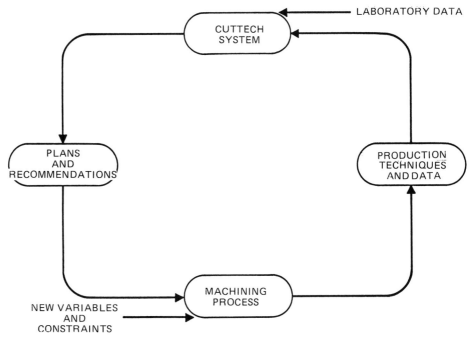

Figure 5.29 The Information Loop in Operations Planning

experience and proven techniques in their field. One application area of knowledge-based systems in FMS is operation planning in machining [12, 2–11].

In machining operations, planning includes the selection of cutting tool, cut sequence, cutting conditions, tool replacement strategy, and so on needed to produce a single feature (e.g., a hole, slot, or bevel) on a part. An information loop of a knowledge-based system application is shown in Figure 5.29 [12, 2–12]. The knowledge-based system called CUTTECH captures metal-cutting technology and data for use in recommending productive and economical tools and cutting parameters for machining. The system's functions include selecting cutting tools, cut sequences, and speeds and feeds for a user-defined part feature to be machined.

The CUTTECH shown in Figure 5.29 acts as both a knowledge source and a knowledge collection point in a machining facility. At the center of the information loop is the actual machining process. The machining process generates real productive data as metal cutting takes place, and CUTTECH provides input to the machining process in the form of an operation plan. The machining process in turn generates new techniques and data as plans are modified to meet actual production requirements. These new techniques and data are documented, along with laboratory data, and become inputs to future production processes by their incorporation into the CUTTECH system.

Unlike many other AI planning systems, CUTTECH adds intelligence to a core of machinability data rather than adding practicality to abstract geometrical concepts for machining. That is, CUTTECH builds basic machining data, increasing its value by using it as input for knowledge-based rules. The value of this system

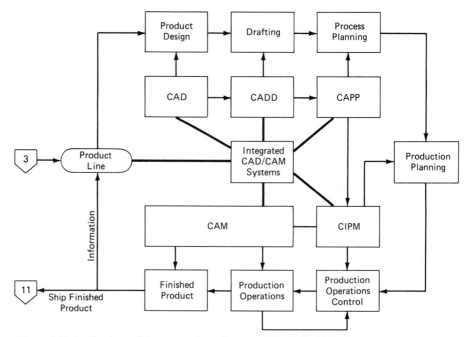

Figure 5.30 Applications of Computers in an Integrated CAD/CAM Environment

is underscored by the fact that, as NC machining continues to become the more popular method for batch manufacturing, the ranks of experienced conventional machinists are declining through promotion and retirement. As a result, the loss of shop floor expertise mandates the development of computerized systems to capture machining knowledge.

An FMS Scenario

Let us begin a production process by using Figure 5.30 as an example. This FMS is a comprehensive package. For example, it involves the use of such automated manufacturing systems (CIMS) as the following:

- CAD (computer-aided design)
- GT (group technology)
- CAPP (computer-aided process planning)
- CAPM (computer-aided production management)
- CAP (computer-aided production)

Computer-aided production includes six modules: fabrication, assembly, test and inspection, material handling and transfer, data communications, and supervisory controls. These modules are integrated to form the thrust of this FMS. Elements of this FMS production process evolve in a closed loop—from product design

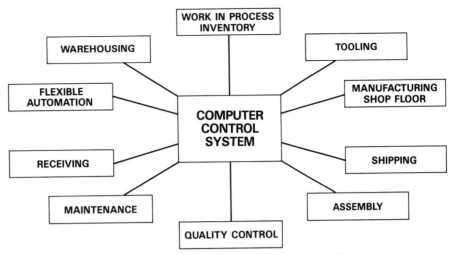

Figure 5.31 Typical Users of a Production System (Real-Time Factory Control)

and refinement to delivery and service by use of an intelligent database system, which through control techniques, permits data and values to be added at various points in the production cycle to meet the needs of specific operations.

Product Design and Refinement

The production process begins when product design receives a decision-to-manufacture (or WAD) authorization. Using the philosophy of "Design or refine only what is necessary" avoids reinventing the wheel. As a result, the database containing preliminary design information that was developed during the early stage of product development is used. The process of designing or refining the product is carried out by interactive communication between designers and computers. Typical users of a CAD/CAM production system in a real-time environment are illustrated in Figure 5.31.

The designer supplies the design concepts and does the creative work. The computer supplies real-time data from concurrent operations in other functions, stored production data, standardized information, and then it carries out the design calculations. During the design process, the computer constantly calls for information from CAPP and CAM. It takes into account information on manufacturing costs and capabilities required to produce on the system's production machines for each feature the designer conceives. Materials are selected and parts are detailed on a CRT terminal. Dimensions are applied to sketches, along with suitable tolerances and clearances. These designed tolerances and clearances are verified automatically.

A finite element model is quickly produced from the data for parts that in the final product will be subjected to stress. This model is sent over communication lines to a larger computer. The analyzed data are displayed graphically to the designer on a color terminal. Areas of greatest stress are highlighted to suggest how to redesign or reinforce the part.

The computer also uses a designer's information to find a design that satisfies the product requirements and that can be produced at minimum cost with maximum efficiency while meeting all requirements. The approved design is released to EDC as common product data for use by various applications. Data from the product design model are enhanced by other applications to meet their requirements. Certain additional elements of information are carried along with the design data to these applications.

Process and Production Planning

The feasibility of producing the design economically is verified by manufacturing engineering. Because the data are stored in an integrated database system, the ability to manufacture a part is verified during the design phase. In this case, the computer facilitates the ability of various functions to work together, to reduce costs, and to increase productivity.

Both GT and CAPP assist manufacturing engineering with the product translation process by extending the product descriptions already in the system. Translation provides the initiating actions for the production processes. Through the use of computers, production standards, and methods, process plans are developed and prepared. The data generated by CAD is used to develop NC tool paths, to generate NC tapes, and to produce parts directly during the production phase. Tooling to fabricate the part is simulated and evaluated by the computer. The designer verifies the operation of the tool robots by observing a simulated performance on a CRT.

Moving more to the production side of the house and extending the product description to generate production plans is an obvious benefit in the elimination of redundant data. Values are added to the data to provide production plans, schedules, budgets, job tracking, shop loading and control, overwriting, procurement, inventory control, and the like. Because the product data are stored in the integrated database system, many planning operations can be performed concurrently as well as serially with other functions.

The Production Process

The planning information is used to drive and control automated machines and equipment at the various FMS workstations. The workstations are linked together by an automated materials handling system consisting of subsystems such as conveyors, cranes, AVG, shuttles, and tow carts. A central computer monitors, controls, and schedules the entire operation.

An FMS builds the product to order. The plant host computer runs a scheduling algorithm that compares order intake and requirements for product parts with part inventories and machine capacities. A production schedule is generated and made available to the cell controller for implementation. This production schedule of the parts and options to manufacture may be for a day or a shift. The automated machines set themselves up, automatically handle parts, select their own tooling, and carry out a variety of fabrication processes of the removal, plastic flow, and consolidation kinds, including assembly of the product.

The cell controller makes available to the plant host computer production information (including product counts and manufacturing time) on demand or on a scheduled basis. The process is continuously monitored by the computer and adjusted for efficiency, quality, and cost. Raw materials and completed parts are moved and stored by automatic systems for materials handling and warehousing. The cost data are saved in the database system for use in the bidding for and construction of similar components. Control information is generated from manufacturing data and factory layout information. At the end of each shift, statistical quality control (SQC) and tool fixture management computers make available to the plant host computer information such as reject rates and tools used in order to evaluate the cost of scrap and new tooling.

Management Information Reports

Cathode ray tubes (CRT) are strategically located around the plant. This gives management an opportunity to make extensive use of the integrated database system. The plant host computer generates periodic reports so that managing the entire factory environment is possible. Status reports and shift reports are transmitted from the factory floor to the host computer. They contain information on material and energy usage, inventory status, and other information helpful to the management and accounting functions of the plant.

Advantages and Disadvantages of FMS

Some advantages of FMS are the following:

- Reduced manufacturing lead time
- Reduced in-process inventory
- Improved product quality
- Reduction in product cost
- Improved use of manpower, equipment, and facilities
- Production of families of workparts
- Better management control
- Random launching of workparts into the system

Some disadvantages of FMS are the following:

- High cost and complexity
- Poor system reliability
- Inadequately integrated database system
- Replacement of human resources
- Increased maintenance costs

Exercises

1. Define (a) AI, (b) ES, (c) GT, (d) FMS.

2. Explain the difference between AI, ES, and knowledge-based systems.

3. Explain the difference, if any, between CAP and CAM.

4. Explain the meaning of CAD/CAM.

5. Explain the difference, if any, between CAD/CAM and ICAM.

6. (a) What is the purpose of the ICAM program? (b) List some advantages of the program. (c) List some disadvantages of the program.

7. What are the ICAM program thrusts?

8. Define CAM–I. What is the difference, if any, between CAM–I and ICAM?

9. (a) What are the Big Three in CAD/CAM? (b) Define the role of each in CAD/CAM.

10. Explain the difference, if any, between a common database in CAD/CAM and an integrated CAD/CAM database system.

11. What is a distributed database system?

12. Discuss the role of data communication in CAD/CAM and the automated factory.

13. Explain (a) distributed control, (b) supervisory control, (c) LAN, (d) MAP.

14. Explain the difference, if any, between automated process planning and computer-aided process planning.

15. Explain the difference between a variant and a generative process planning system.

16. List and define three popular GT classification and coding systems.

17. Explain the difference between FMS and an automated factory.

Self-Study Test

1. An integrated CAD/CAM database system is often referred to as the manufacturing database.
   True   False

2. CAD/CAM may be defined as a closed loop system reflecting all activities included in the production cycle.
   True   False

3. The primary difference between broadband and baseband LANS is:

  1. LAN protocols
  2. The method of signal generation
  3. The area LAN encompasses
  4. Regulations of the FCC and/or state public utility commission

4. The term _____ is commonly used today to describe networks.

  1. Protocols
  2. Architecture
  3. ISO

5. Typical CAM applications for manufacturing planning are:

  1. Computer-aided line balancing, process control, cost estimating, CAPP, and process monitoring
  2. Computerized machinability data systems, computer-assisted NC part programming development of work standards, and production and inventory planning
  3. Shop floor control, process control, production and inventory planning, and cost estimating
  4. All the above.
  5. None of the above.

6. Group technology is a manufacturing philosophy in which:

  1. Similar parts are identified and grouped together to take advantage of their similarities in production and design.
  2. Group technology may be considered as the "glue" that holds production and design together
  3. Similar parts are arranged into part families
  4. All the above.
  5. None of the above.

7. A list of the sequence of moves between operations or work centers is called a:

  1. Routing sheet
  2. Process sheet
  3. Production schedule
  4. Work-in-process sheet
  5. Capacity requirements planning

8. Machinability data systems select cutting speed and feed rate on the basis of the following characteristic(s) of the operation.

  1. Type of machining operation
  2. Machine tool and cutting tool

3. Workpart and operating parameters other than feed speed

4. All the above.

5. None of the above.

9. CIM includes:

1. All the engineering functions of CAD/CAM

2. Business functions

3. Design, manufacturing planning, and manufacturing control

4. Marketing and product support

5. All the above.

10. _____ refers to production management techniques for collecting data from factory operations and using the data to help control production and inventory in the factory.

1. Process control

2. Shop floor control

3. Manufacturing control

4. Computerized factory data collection and computer process monitoring techniques

5. Items 2 and 4 above

11. _____ is defined as the effective use of computer technology in the planning, management, and control of the manufacturing function.

1. CAE

2. CAD

3. CAM

4. CAPP

5. CAD/CAM

12. CAD/CAM involves the use of a digital computer to accomplish certain functions in:

1. Design

2. Process Planning

3. Production

4. Items 1 and 3

5. All the above.

13. Flexible manufacturing systems incorporate automation concepts such as _____ into a single system, station, or cell.

1. Automatic material handling between machines; numerical control machine tools and CNC; and group technology

2. Computer control over the materials handling systems and machine tools (DNC); group technology; and numerical control machine tools and CNC

3. Equipment maintenance and repair; tool changing and setting; and numerical controls

4. Items 1 and 2 above

5. All the above

14. The term _____ is a basic term that defines how network components establish communications, exchange data, and terminate communications.

1. Protocol

2. Packet

3. ISO

4. OSI

15. MRP–II is a people system (aided by the computer) as well as a way of doing business. It is a method of top-down scheduling of all the manufacturing resources required in the production of a product.
True   False

16. Adaptive control (AC) has attributes of both feedback systems and optimal control systems.
True   False

17. The material requirements planning (MRP) system starts with the company's business plan, which defines the company's goals and objectives.
True   False

18. Shop floor control (SFC) is concerned with the release of the production orders to the factory, controlling the process of the orders through the various work centers, and acquiring current information on the status of the orders.
True   False

19. Off-line inspection is achieved by performing the inspection procedure during the production operation.
True   False

20. Flexible inspection system (FIS) is related to coordinate measuring machine (CMM) the way FMS is related to computer-aided test (CAT).
True   False

21. The purpose of automated materials handling in a factory is to move raw materials, work in process, finished parts, tools, and supplies from one location to another to facilitate the overall operations of manufacturing.
True   False

22. An AGVS is a materials handling system that is guided along defined pathways in the floor.
True   False

23. A LAN contains four major components: path, interface, protocol control, and data link control.

    True   False

24. Data line controls (DLC) can be described and classified by:

    1. Message format

    2. Line control method

    3. Error handling method

    4. Flow control procedures

    5. All these.

ANSWERS TO SELF-STUDY TEST

1.	True	**6.**	4	**11.**	3	**16.**	True	**21.**	True
2.	True	**7.**	1	**12.**	5	**17.**	False	**22.**	True
3.	2	**8.**	4	**13.**	4	**18.**	True	**23.**	True
4.	3	**9.**	5	**14.**	1	**19.**	False	**24.**	5
5.	4	**10.**	5	**15.**	True	**20.**	False		

References

[1] Blauth, Robert E., "What is CAD/CAM?" Computervision Corporation, Bedford, MA, 1980.

[2] U.S. Air Force, "U.S. Air Force Integrated Computer Aided Manufacturing," *ICAM Program Prospectus.* Wright-Patterson Air Force Base, 1979.

[3] OIR, "Group Technology." *CAD/CAM Integration,* Organization for Industrial Research, Waltham, MA, 1979.

[4] Kinnucan, Paul, "Computer-Aided Manufacturing Aims for Integration." *High Technology,* May–June, 1986.

[5] Weber, Thomas, "The CAD/CAM Data Base . . . The Foundation of CIM." *Commline,* January–February 1985.

[6] ———, "Plant-Wide Communication." *Straight Talk,* No. 4, Allen-Bradley Systems Division, Highland Heights, OH 44143, 1986.

[7] ———, "Plant-Wide Communication." *Industrial Automation Productivity Issues,* Allen-Bradley Systems Division, Highland Heights, OH 44143, 1985.

[8] Waterbury, Robert, "Computer Assisted Process Planning—Key to Cost Savings." *Assembly Engineering,* 1980.

[9] ———, "Chain-Type Code Illustration." *Proceedings of the 5th Annual D CLASS Conference,* BYU, Provo, UT, 1984.

[10] ———, "CAD/CAM Integration Through GT." *Assembly Engineering,* 1980.

[11] _____, "CAD/CAM Paves the Way For The Automated Factory." *Compressed Air,* September 1982.

[12] Barkocy, B. E., and Zdeblick, W. J., "A Knowledge-Based System For Machining Operation Planning." *AUTOFACT 6 Conference Proceedings,* CASA/SME, Dearborn, MI, 1984.

[13] IBM, "Computer-Aided Manufacturing." *Manufacturing Industry Marketing,* White Plains, NY, 1973.

[14] Groover, M. P., and Zimmers, E. W., Jr., *CAD/CAM Computer-Aided Design and Manufacturing.* Prentice-Hall, Englewood Cliffs, NJ, 1984.

[15] Drozda, T. J., Stranaham, J. D., and Farr, G., *Flexible Manufacturing Systems,* 2nd ed. Society of Manufacturing Engineering, Dearborn, MI, 1988.

[16] Coyle, L. W., and Dworschak, J. R., "Distributed Numerical Control—A Tool For Factory Automation." *AUTOFACT 6 Conference Proceedings,* CASA/SME, Dearborn, MI, 1984.

[17] Hyer, N. L., and Wemmerlov, U., "Group Technology and Productivity." *Harvard Business Review,* July-August 1984.

[18] Vrba, J. A., "CAM For the 80's—Distributed Systems Using Local Area Networks." General Electric Company, Oak Ridge, TN, pp. 10–14.

[19] McDonald, J. L., and Hastings, W. F., "Selecting and Justifying CAD/CAM." *Assembly Engineering,* April 1983.

[20] Bishop, H. M., "Road to CAD/CAM." *CAD/CAM Technology,* Summer, 1983.

[21] Freedam, M. D., "The Automated Factory in the 90's." *CAD/CAM Technology,* Summer, 1983.

[22] Gould, L., "Computers Run The Factory." *Electronics Week,* March 25, 1985.

[23] Groover, M. P., *Automation, Production Systems and Computer-Aided Manufacturing.* Prentice-Hall, Englewood Cliffs, NJ, 1980.

[24] Davis, G. Brent, "CAM: A Key to Improving Productivity." *Modern Machine Shop,* September 1980.

[25] Meister, Ann E., "Ironing Out the Rough Spots Between CAD and CAM." *CAD/CAM Technology,* Fall, 1984.

6

Computer-Aided Business

Overview

Information technology is having a crucial impact on both business and industry, changing corporate structures, products, and processes. Information technology is also having a decisive effect on the competitiveness of industries. Recent advances have caused the merging of previously unrelated areas in the corporation as well as the integration of corporate information systems more closely with external systems.

Rapid advances in mini- and microtechnology have provided solutions for a wide range of management problems. The concept of a distributed information processing system offers attractive alternatives to centralized systems, not only in the office but, more important, in the production areas of the factories, where critical data can be captured and analyzed by management as an integral part of the manufacturing or assembly process.

With recent advances, management's expectations of the benefits of computer technology in the future may be greater and more optimistic than were past expectations. Technology now exists for companies to do most operations that they want in distributed data processing, information processing, and communication. Therefore, the understanding, involvement, and cooperation of management at all levels are needed to realize the potential that is offered.

As indicated in earlier chapters, computer-integrated manufacturing (CIM) is a systems integration concept, not a technology. That is, it is a management philosophy that advocates the integration of all technologies in the manufacturing environment. It looks at a company's production resources as a single system, and it looks at defining, funding, managing, and coordinating all improvement projects in terms of how they effect the entire system.

In addition to product resources, other aspects of a company are a part of the manufacturing environment. The business systems (components) such as marketing, sales forecasting, scheduling, shop floor control, order entry and purchasing, and manufacturing resource planning are integrated into CIM (Figure 1.19).

The integration of the business components into the manufacturing environment can be viewed as the computer aiding the business segment, which as was said in Chapter 1, is called computer-aided business (CAB). With the support of business components in the manufacturing environment, the total company resources are brought to the support of the product production area. In the past, the business components have indirectly assisted the production area, but they assisted *outside* the circle of production. Now, however, the business components are being integrated into the manufacturing environment. With the use of computers to support the business components, we see a logical reference for CAB. This chapter discusses the business components that are integrated into the manufacturing area as support for the product production environment.

Marketing Production Support

Marketing Planning Overview

Industries are traditionally organized according to the function of their parts. Such functions are marketing, finance, and production. All kinds of industries have a finance function, and most of them have a marketing function. Only those organizations that produce the products they sell, however, have a formal production function, called manufacturing.

The financial function is concerned with the money flow, and the traditional manufacturing and marketing functions are concerned with the material flow (material includes both products and services). The marketing function determines what products and services should flow from the firm to its customers, and the manufacturing function creates or provides the products and services.

Many companies have moved toward a marketing orientation. This move means that the entire organization is dedicated to the objective of marketing, which is the satisfaction of customer wants and needs at a profit. This objective is called the *marketing concept.*

In obtaining the objective of the marketing concept, a marketing plan is implemented. A marketing plan is a long-range plan to assist management in controlling and monitoring the marketing function. With a marketing planning system established, the system aids management in structuring and maintaining an annual plan to ensure that directives are obtained. The system assists management in evaluating alternative marketing plans and coordinating marketing plans with distribution, production, research and development, and finance. The collective result is dependent upon the ability of the marketing function to deliver the forecast level of revenues, through a complex scenario of constantly changing product mix, prices, and advertising and distribution expenses.

The Mix in Marketing

One of the marketing manager's objectives is to develop strategies that enable the resources listed in the previous paragraph to be used in marketing the firm's products and services. The strategies consist of a mixture of ingredients called the marketing mix.

The *marketing mix* is the package of products and services presented to a prospective customer as a means of satisfying the customer's needs and wants. The marketing mix consists of products, promotion, place, and price—the four Ps. *Product* is what the customer buys to satisfy the perceived needs or wants. *Promotion* is the means of encouraging the sale of the product, including advertising and personal selling. *Place* is physically distributing the product to the customer. Ingredients in "place" include transportation, storage, and distribution for the manufacturing wholesaling, and retailing of the product or service. *Price* is what the customer must pay for the service or product, including discounts and bonuses.

Place Marketing Mix Subsystem

The method a company uses to make products available to the customers is considered the "place" ingredient in the marketing mix. Place decisions are of two basic categories.

Establishment of channel systems. Products and services are made available to customers through channel systems. The company must decide what other companies will be members of the channel and what resource linkages must be established to facilitate the product flow. Selection of the channel members is an informal process. The process can be made more formal by gathering data describing potential channel members and processing the data on the computer according to preestablished criteria.

In establishing the necessary resource linkages, attention must be paid to the flow of physical and information resources. The channel member should demonstrate ability to establish the necessary transportation, storage, and data processing procedures. The information flow is not easy to achieve because the information should flow in both directions—toward the channel member and toward the manufacturer.

Each channel member must know the details of the product flow at each point in the channel. The manufacturer should know the rate wholesalers are paying for the product, the rate retailers are paying to wholesalers, and the rate consumers are paying to retailers. Manufacturers also need information after the physical product flow occurs, whereas the wholesaler and the retailer need information before the flow begins. Feed-forward information from the manufacturer to the channel members can include announcements of new products, sales and promotion aids, and forecasts of demand. If participants in the channel system realize the value of the information flow and the improved performance it offers, then an efficient inter-firm information system is possible.

Information flow can be achieved in several ways. Sales representatives of the channel members prepare written reports and communicate information by word of mouth. Information is frequently transmitted by a data communication network. A channel system that permits information to flow freely among the companies (the channel members) provides an edge over a competitor who does not have that capability.

Performance of the distribution functions. When the channel members have been selected and arrangements made for the flow, the channels can go into operation. This area of marketing is where the computer has the potential for the greatest effectiveness because the problems of physical distribution (i.e., logistics) are essentially structure and determination. In addition, such variables as shipping and storage costs are known and are measured quantitatively. The computer can be used to keep the distribution costs to a minimum.

Promotion Subsystem

Full use of the power of a computerized information system in personal selling, advertising, and promotion has been difficult to achieve. These areas are important ingredients in the marketing mix. Sales reporting systems have been used for years to provide a record of past performance. Unfortunately, few reporting systems exist in advertising.

Advertising is more of an art than a science. Creativity is a big part of advertising, and marketers actually know little about why some ads encourage purchases and others do not. The computer has been used in advertising to aid in deciding how much money should be allocated and how to allocate money by relating advertising to such indicators as population trends, birth rates, income levels, and interest rates. The economic models generally use "canned" statistical software and databases of economic information.

Generally speaking, however, it has been difficult to apply the computer to any phase of advertising. One form of advertising, public relations (PR), has been an especially difficult field in which to use computers to advantage. Public relations is the conveying of a promotional message by a third party—such as newspaper columns that review movies and TV programs.

Decisions related to personal selling are more structured than those of advertising or public relations because they do not depend as much on whims of the public. Many industries use computer-based planning techniques to plan sales activities for the upcoming year. These plans are based on a sales forecast, which provides a basis for determining recruiting and training needs. Such plans are used to develop personnel schedules, selection of new hires, and training programs for the year to meet annual sales objectives.

Sales managers use electronic spreadsheets for both planning and controlling. The spreadsheets make two main contributions, as follows:

1. They enable the manager to play the what-if game in making various critical decisions.
2. They can bring to light some rather subtle characteristics of the data that might go unnoticed.

Price Subsystem

Marketers tend to follow two approaches in pricing policies. One is a cost-based pricing policy, by which costs are determined and the desired markup is added to arrive at a price. This approach is considered cautious. A second, less cautious approach is a demand-based policy, by which a price compatible with the value that the buyer places on the product or service is established. Higher profits can be made from a sale using this approach than from one made using the more conservative cost-based policy. When the cost-based policy is used, the marketer is constrained from setting the price too high because of competition.

An information system can support management in both pricing policies. With the cost-based approach, the system can provide accurate cost accounting data on which to base a decision. With the demand-based approach, the system enables the manager to engage in what-if modeling to locate the price level that maximizes profit and restrains competitive activity.

Mathematical models can be used to simulate the effects of the firm's pricing strategy for profits. The models should consider both internal and external influences, which include the following:

External Influences

• National economy
• Seasonal demand
• Competitor's pricing strategy
• Competitor's marketing budget

Internal Influences

• Plant capacity
• Raw materials inventory
• Finished goods inventory
• Marketing budget

Data representing each of these influences are entered into a model. The data represent the scenario that will produce a likely output. Either a procedural language or a spreadsheet can be used to create the model, but one should realize that a model is only as good as the mathematics and the data on which it is built.

Computer-Integrated Production Management Systems (CIPMS)

Computer-integrated production management has traditionally been called production planning and control. The computer has become a powerful tool to help accomplish processing and routine decisionmaking chores in production planning and control that have previously been done by humans alone. Computerized information systems are designed to integrate the various functions of production planning and control to reduce problems during the planning and execution phases of the manufacturing cycle. Such problems typically have included the following:

1. *Plant Capacity.* Because of lack of labor and equipment, production falls behind schedule. Excessive overtime, delay in meeting delivery schedules, customer complaints, and back-ordering result from this problem.

2. *Suboptimal Production Scheduling.* The wrong jobs are scheduled because of lack of clear priorities, and inefficient scheduling rules. As a result, production runs are interrupted, machine setups increase, and jobs that are on schedule fall behind.

3. *Long Manufacturing Lead Times.* Production planners allow extra time to compensate for problems of labor and equipment, or of inefficient scheduling. As a result, the shop becomes overloaded, order priorities become confused, and long manufacturing lead times exist.

4. *Inefficient Inventory Control.* Total inventories are too high for raw materials, work in progress, and finished products. At the same time, stockouts occur on individual items needed for production. High total inventories mean high carrying costs, and raw materials stockouts mean delays in meeting production schedules.

5. *Low Use of Work Center.* This problem results from poor scheduling, equipment breakdowns, strikes, and reduced demand for products, over some of which management has little or no control.

6. *Process Planning Not Followed.* With this problem, regular planned routing is superceded by an ad hoc process sequence. The problem is caused by bottlenecks at work centers in the planned sequence. The results are longer setups, improper tooling, and less efficient processing.

7. *Errors in Engineering and Manufacturing Records.* Bills of materials are not current, route sheets are not up to date with the latest engineering changes, inventory records are inaccurate, and production piece counts are incorrect.

8. *Quality Problems.* Quality defects are encountered in manufacturing components and assembled products, resulting in rework or scrapped parts and consequent delays in shipping.

The computer integration of production planning and control functions is to increase efficiency of shop operations by better production schedule planning and balancing the production workload to production capacity. Improved customer service is the result of these efforts because of the resulting reduced time from customer order to delivery and because the right products to meet customer requirements have been produced. Investment in raw material inventory is reduced because of improved planning and control of the facilities. Better use of the facilities gives higher productivity and improved quality control because of the continuous monitoring and feedback to shop operations.

The typical functions of CIPM include forecasting, production planning, master scheduling, materials requirement planning (MRP), capacity planning, engineering and manufacturing database management, shop floor controls, shipping and inventory control, purchasing, and process controls. Computerized systems have been developed to perform these functions, but the functions themselves have remained unchanged in many current manufacturing industries. The more significant changes, however, have occurred in the organization and execution of production planning and control through MRP, capacity planning, and shop floor control.

Basically, CIPM can be viewed as consisting of five major subsystems:

1. *Master Production Schedule Planning.* This subsystem helps to develop a feasible master production schedule for effective management of production inventory. This subsystem is vital in determining such long-range resource requirements as cash, shop capacity, and long lead time on raw material requirements.

2. *Production Inventory Management.* This subsystem explores the master production schedule to pinpoint specific raw materials requirements. It then plans production lots based on forecasted job orders.

3. *Plant Shop Scheduling and Loading.* This subsystem plans the detailed capacity requirements by assigning starting dates to each production lot and then analyzing the shop production load. When the schedule is firm, the system checks inventory levels, allocates the raw materials and facilities, and releases orders to the shop.

4. *Shop Monitoring and Control.* This subsystem provides the basis for continuous information feedback to shop management by capturing and controlling all labor, production, and maintenance transaction data such as labor reporting, machine downtime reporting, and production counts.

5. *Plant Maintenance.* This subsystem assists management in manpower planning, work order dispatching, maintenance costing, and preventive maintenance scheduling. The objective of this subsystem is to improve the overall efficiency of the job shop while holding maintenance, labor, and material costs to a minimum.

The remainder of the chapter addresses functions having the most impact in CIPMS.

Control of the Shop Floor

Control of the shop floor is a system of monitoring the status of production activity in the plant and reporting the status to management so that effective control is established. The computer is used as a shop floor information system to provide accurate, timely shop data to manufacturing control systems involved in scheduling, inventory, or quality. Efficient operation of a manufacturing shop operation depends on information to control daily operation, to meet schedules, to control inventory, to acquire needed resources, and to maintain quality.

Production managers must acquire up-to-date information on the progress of orders in the factory and make use of that information to control factory operations. This problem is addressed by a shop floor control system.

A control system monitors and controls production equipment and captures status information to provide notification to production management of an out-of-control situation. The system provides direct computer control of process equipment and makes the necessary adjustments automatically.

A shop floor control system should be designed to accomplish the following functions:

1. Priority controlling and assigning of shop orders.
2. Maintenance of information on work in progress for MRP.
3. Monitoring shop order status information.
4. Providing production output data for capacity control purposes.

The control system generates information for individuals to use in making good decisions on effective factory management and execution of the master schedule.

A shop floor control system consists of three steps:

1. *An order release* provides the documentation for an order as it proceeds through the shop. The documentation includes route sheet, material requisition, job cards, more labor tickets, and parts list.
2. *An order schedule* makes assignments of orders to machines in the factory. The priority of a job order is determined by its due date.
3. *The order progress* report provides data related to work in process, shop order status, and capacity control. Data are collected from the shop floor, and reports are generated to assist production management.

Order Entry and Purchasing

An order entry system enters customer purchase orders into the system. The system can also reject customer orders if there are such problems as poor credit. If an order is accepted, it is processed by the inventory system where records of the company's physical inventory are updated.

If inventory is on hand to fill customer orders, invoices are prepared by the billing system, and the accounts receivable department is advised of the transaction.

Accounts receivable handles the collection of money by sending statements, or invoices, to the customer.

If the inventory is not on hand to fill customer orders, the inventory system informs the purchasing system to buy the needed inventory (assuming that the purchase must be made from a vendor instead of producing it in-house). A purchase order is prepared and accounts payable is advised of the transaction. The receiving system informs the inventory system when the stock has been received so that inventory can update the inventory records. Accounts payable is also informed so the vendor can be paid. Figure 6.1 shows the flow of the order entry system.

Figure 6.1 Order Entry System

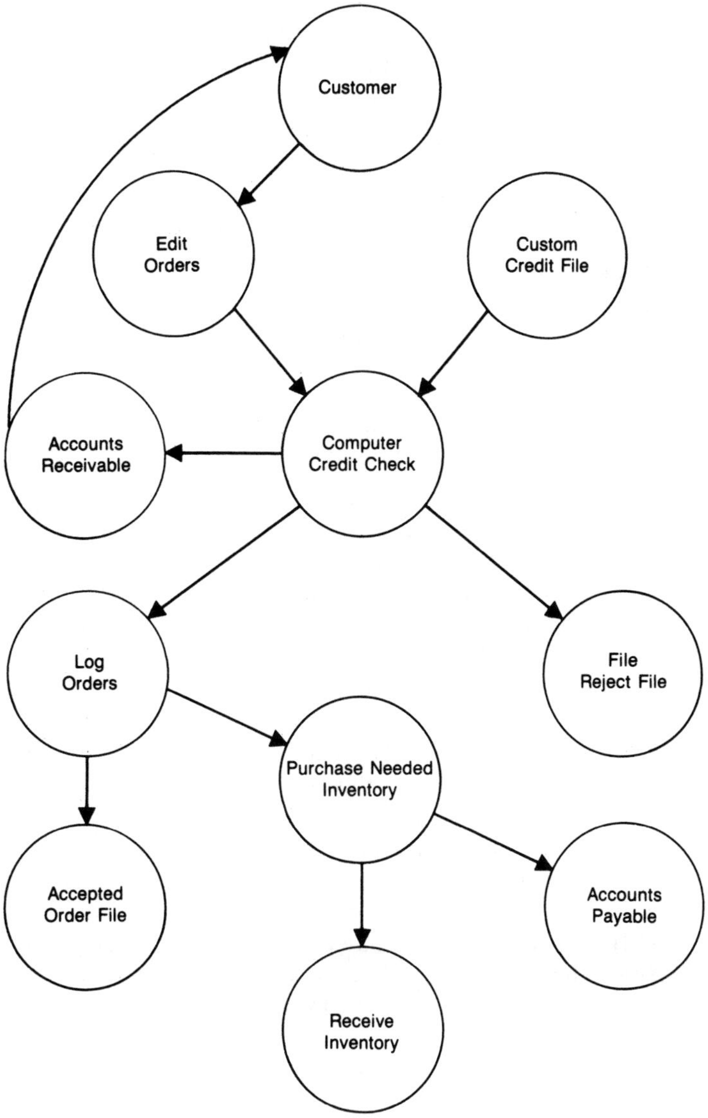

Purchasing

The company will manufacture some of the inventory components and purchase others. For the components produced in-house, raw materials are acquired. Ordering the raw materials is the function of the purchasing department. Materials are ordered, and accounts payable is informed of the liability incurred for payment of the bill.

The purchasing activity is triggered by the inventory system when materials must be ordered. Buyers in the purchasing department select vendors for the purchase. Factors in selecting vendors are the quality of their product, their ability to meet delivery dates, and the prices they charge.

Purchasing is the least computerized process in the order entry and purchasing system. The computer provides a supporting role by signaling when it is time to order or reorder, providing vendor information, and printing the invoices.

One reason the computer has not been used extensively in the vendor selection process is that procurement changes frequently. Vendors come and go, prices change, vendors add and delete products. Because the buyer must review the scenario at the time of each procurement, little is left for the computer to do.

Material Requirements Planning

Joseph Orlicky devised an elemental materials requirement planning (MRP) package in the early 1960s, thus originating the MRP concept. The first commercial MRP package was the production information and control system (PICS) from IBM, which became available in the late 1960s.

Materials requirement planning is a technique of managing production inventories that takes into account the specific timing of the requirements. It is a schedule of material required in each period by the firm's production schedule. The production schedule is determined by the sales forecast provided by the marketing information system. An MRPS is a positive approach to materials management—anticipating material needs and planning the acquisition.

An MRPS interacts with two systems: production scheduling and capacity requirements planning. It was not until the early 1970s that the capacity requirements planning feature was added, along with feedback from the shop floor to produce a closed-loop MRPS. The status of both the capacity and the performance of the plant in terms of the schedule could be updated as the production occurred. Data collection terminals facilitated the feedback from the shop floor.

When an MRPS interacts with two production systems—scheduling and capacity requirements planning—the following steps are taken:

Step 1. The sales forecast is used to create a master production schedule (MPS). The sales forecast identifies the quantities of the various finished goods to be sold. The period covered by the schedule can be a year or more. The schedule should be able to accommodate the longest vendor lead time plus the time needed to produce the item when all the materials are available.

Step 2. The material requirements planning system uses the MPS to determine the types and quantities of raw materials needed to produce the finished goods. The determination is made by "exploding" the bill of materials. The bill of materials is simply a list of all raw materials and their quantities needed to produce one item of finished goods. Exploding is accomplished by multiplying the number of units of finished goods to be produced by the quantities of needed raw materials.

The total quantities of raw materials needed are called the gross requirements. The raw material inventory file is checked, and the quantities subtracted from the gross requirements. The balance, identifying the materials that must be purchased, is called net requirements. The net requirements are allocated to the different periods to reflect vendor lead times and specific steps during the production process when the materials will be needed. An MRPS has the ability to schedule capacity on an overtime basis if management wants to produce more than the regular capacity permitted by the system. When the capacity constraints have been satisfied, the MPS in Step 1 is changed frequently to reflect changes in the business. The ability to quickly and easily reschedule is a big advantage of advanced MRP systems.

The outputs required by the MRPS include (1) a planned order schedule that lists needed quantities of each material by period and is used by buyers to negotiate with vendors; (2) order releases that are authorizations to produce the products on the order schedule; and (3) changes to scheduled orders that reflect cancelled orders or modified order quantities.

Optional outputs include (1) exception reports, which flag items requiring management attention; (2) performance reports, which indicate how well the system is performing in terms of stockouts and backorders; and (3) planning reports, which are used by manufacturing management for future inventory planning.

A well-designed and well-managed MRP has many advantages, among which are the following:

- *Reduction in Inventory.* An MRPS mainly affects raw materials, purchased components, and work-in-process inventories. Users reported a 30 to 50 percent reduction of work-in-process.
- *Improved Customer Service.* Some MRP proponents have reported that late orders are reduced by 90 percent.
- *Response Time.* There is a quicker response to changes in demand and in the master schedule.
- *Greater Productivity.* Reports were that productivity can be increased by 5 to 30 percent through MRP.
- *Reduced Setup and Product Changeover Costs.* Few changes because of accurate engineering and production data base.
- *Better Use of Machines.* Application computerized process planning and group technology
- *Increased Sales and Reductions in Sales Price.* Reduction in the manufacturing cycle time activities due to systems integration improved product quality and reliability.

Manufacturing Resource Planning

The previous section provided an overview of MRP as a guide to understanding the concept, which was a hot topic in the manufacturing world during the 1970s. Many companies tried it. Most of them were large companies because the software was available for mainframes and minis only. Companies saw the MRPS as a way to achieve goals of reduced costs, increased efficiency, and improved responsiveness to changes in consumer demand as reflected in the sales forecast.

Not all users achieved their goals with MRP, although those that did not rely on MRP alone did achieve their goals. Those companies made changes throughout the company, beginning at the top-management level and extending to their customers and vendors. The company-wide view of MRPS has been named manufacturing resource planning (MRP–II), and it is a formal system for managing a manufacturing business.

The key to using MRP–II successfully is to link each of the plans, priorities, and actions of manufacturing, marketing, finance, engineering, and personnel to the closed-loop production and inventory control system. Because it is a closed-loop, information flows back to each function to accomplish management's goals and objectives for the company. Thus, Figure 6.2 shows the requirements for MRP II. Here is how MRP II works:

Figure 6.2 MRP II Requirements

Business Plan

Production Plan

Master Production Schedule

Material Requirements Plan

- Complete/Accurate Bill of Material

- Complete/Accurate Inventory

Capacity Requirements Plan

- Complete/Accurate Manufacturing Routings

Procurement Plan/Execution

Manufacturing Plan/Execution

Feedback

Financial Control

Performance Measurement

1. It establishes a specific planning order.
   - The business plan reflects the market plan and forecasts.
   - The production plan supports the business plan.
   - The master production schedule supports the production plan.
   - Both the production plan and the master production schedule are balanced against material and available capacity.
   - The plan for material requirements is driven by the master production schedule, which generates procurement and manufacturing plans.
   - Manufacturing plan balanced against capacity requirements plan.
   - Each step in the planning process is positively resolved before proceeding to the next step.
2. It establishes a feedback closed-loop process.
   - The procurement plan is executed and maintained through feedback.
   - The manufacturing plan is executed and maintained through feedback.
   - Problems are resolved at the lowest possible level, and plans are updated to reflect current priorities.
   - The emphasis is on seeking resolution by going back one step at a time—not starting at the top.
3. The impact of change is assessed through simulation.
   - Change at any level of the planning process explodes downward and sufficiently into the future for impact on subsequent plans to be determined.

Evolution of MRP to MRP–II

Four steps can be identified in the evolution of MRP.

1. *An Improved Ordering Method.* The initial use of the computer was to perform the calculations of requirements planning. Computerized MRP systems are an improvement in the ordering of raw materials and components because of the speed and accuracy with which the requirements planning task can be performed.
2. *Priority Planning.* The MRPS began to incorporate priority planning into computations to generate schedules and requirements that could be accomplished by the factory. An MRPS determines not only what materials should be ordered but also when those materials will be required (see Figure 6.2). Priority planning provides a means of dealing with rush jobs by increasing their priorities, in addition to helping to remove from an expedited status those jobs that no longer have priority.
3. *Closed-Loop MRP System.* The closed-loop MRP system is an improvement over the MRPS discussed in item 2 above because it not only plans the priorities but also provides feedback relative to executing the priority plan. A closed-loop MRP system means that the various functions in production planning and control (capacity planning, inventory, management and shop floor control, and MRP) have been integrated into a single system. Feedback is also received from

vendors and the production shop when problems arise in implementing the production plan.

4. *The MRP–II System.* The MRP–II involves a linkup between the closed-loop MRP system and the financial systems of the company. Manufacturing resource planning (MRP-II) is the name for this combination.

An MRP–II system has two basic characteristics (see Figure 6.3) beyond a closed-loop MRP system:

1. *An MRP–II system is an operational and financial system.* This aspect makes MRPS–II a company-wide system that is concerned with all facets of the business, including sales, production, engineering, inventories, and cash flow. The operations of the individual departments are reduced to financial data. This common base provides the company's management with the information needed to manage it successfully.

2. *An MRP–II system is a simulator.* As a simulator, an MRP–II system is intended to answer what-if questions. The simulator can be used to simulate the probable outcomes of alternative production plans and management decisions that are under consideration.

Figure 6.3 Function for MRP and MRP–II

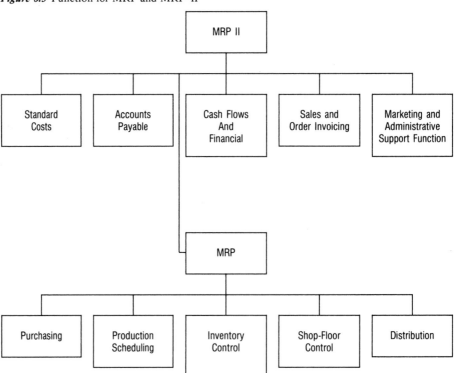

Successful Implementation of an MRP–II System

To implement a successful MRP–II system, top management can take the following actions [6]:

1. Recognize that MRP–II is a disciplined way of conducting the business of the company, and learn the concept and technology firsthand.
2. Appoint a task force led by an executive.
3. Provide the necessary resources by assigning people full time to MRP–II and making MRP–II a top priority for all managers.
4. Develop a formal implementation plan covering about two years into the future before work starts.
5. Insist that vendor-supplied MRP software be used, because it will usually work with little or no modification.
6. Make sure everyone involved receives an education on what MRP can do and their role in it.
7. Require that marketing, finance, manufacturing, and personnel jointly engage in the implementation process.
8. Be patient. Some results can be expected in the first years before the system is completely implemented, but others will take a long time to be fully realized.

No two companies using the same software and hardware will have the same success. One project may fail and the other may succeed. Anderson and Schroeder [6] believe that top management is the determining factor in achieving success. Their commitment and willingness to make changes throughout the company are the keys to a smoothly functioning system.

Software for MRP–II

A wide variety of MRP–II software is on the market to service those firms that do not create their own. Customized MRP–II software can be very expensive; therefore, many firms use packaged software.

As early as 1986 prices of micro MRP software ranged from $10,000 to $36,000, and most were written to interface with MS–DOS or PC–DOS. Today, prices of these systems have dropped tremendously with an increase in efficiency and reliability. A wide variety of micros can be used. In some cases, several micros are networked so that multiple users can access the system.

The most popular MRP package is MAPICS (Manufacturing Accounting and Production Information Control System), which is marketed by IBM to run on the System/34 mini. An estimated 35 percent of the installed MRP systems use MAPICS. The Gulf and Western Manufacturing Company of Southfield, Minnesota, has installed more than 10 MAPICS since 1980 at an average cost of $300,000. They have also installed four of IBM's mainframe MRP systems called COPICS (Communications Oriented Production Information and Control System) at a cost of about $500,000 each.

The average MRP installation cost is about $400,000—73 percent for hardware and 27 percent for software. The software cost appears low, but it is increasing due to high labor costs. However, today with the increased use of AI technology, the software cost is on the decline. The appearance of an increase in overall expense is due to the fact that today's MRP–II systems are much more advanced, sophisticated, and have broader applications than those of only a few years ago. The reason for the higher overall expense is the fact that today's MRP–II systems are much more sophisticated than those of only a few years ago.

If a firm wants to use MRP–II but doesn't want to invest in its own hardware and software, it can subscribe to an MRP–II timesharing service. The ASK Computer Systems of Los Altos, California, makes its MANMAN and PLANMAN systems available for about $1,200 for the first terminal and $750 to $1,000 for additional terminals. In addition, a firm pays for the time it is connected to the central computer. Total costs run about $6,000 a month.

The Environmental Impact of MRP–II Systems

Customers benefit from MRP–II by receiving products on time and perhaps at a lower price if the cost savings are passed along. An MRP–II system causes a major change in how the company orders its raw material, although this change can place a strain on its vendors. The system looks into the future and triggers orders far in advance. Because MRP frequently changes the production schedule, it is common for several change orders to be issued to vendors before the material is actually shipped. Vendors are not always equipped to respond to these changes as quickly as necessary. A good way to solve this problem is for vendors also to use an MRP–II system.

Operations Management

Manufacturing managers are concerned with maintaining the steady flow of materials through the plant. They focus primarily on the current year's operation, giving special attention to the current month, week, or day. The manufacturing information system enables manufacturing managers to plan and to prepare for the production process and then to monitor that process to assure that the schedule is met. The system provides a means by which they can view the production operation as it occurs.

Operations management is concerned with the effective use of people, products, and processes in the modern manufacturing industry. This goal is achieved by proper selection and implementation of policies and plans. Consequently, computers make it easier to keep operations management aware of situations regarding orders, finished goods, work in progress, and costs. Such information is of immediate tactical use in production planning, production scheduling, raw material purchasing, delivery date forecasting, and others.

Changes in Operations Environment

The explosion of on-line applications of technology and increased sophistication in operating systems in the past decade has taken what was a batch (a group of jobs run on a computer at one time with the same program), job-shop environment with heavy human control and turned it into a process manufacturing shop that is self-scheduled and monitored 24 hours a day. The change in manufacturing workflow has triggered a rethinking of what is appropriate scheduling and what is the definition of adequate service levels.

Management has increasingly recognized that there is no such thing as an ideal management control system or set of performance measures. The trade-off between quality of service, response time of online systems, handling of unexpected jobs, total cost, and ability to meet published schedules on both systems appropriately varies from one organization to another.

Information systems must strike varying balances between efficiency (low-cost production) and effectiveness (flexibility) in responding to an unplanned, uneven flow of requests. Information systems cannot be all things simultaneously to all people but must operate on a set of priorities and trade-offs that stem from corporate strategy. Implementation of this idea has sparked the reorganization of some large information systems into a series of focused, single-service groups, each of which can be managed to achieve quite different service objectives.

Information system technology continues to change. The changes initiate the normal problems of change and new operating procedures, while offering potential benefits of lower costs and new capabilities. Key issues for the operations manager are staff, capacity, telecommunications, proper assessment, assimilation and integration of software, and service emanating from outside the organization.

Senior management must assess the quality of support provided by the information system and involve themselves appropriately, depending on how critical the information is to the overall strategic mission of the organization. The major question that should be addressed is whether the information system as presently organized effectively supports the firm.

Overview of Operations Management

Operations management can be defined as the management of a productive system that transforms inputs into outputs. The operations manager follows the policies set by top management and works cooperatively with other organizational functions such as finance, marketing, and engineering. With some control over the inputs, such as labor, use of capital, facilities, and available technology, the primary concern is the management of the transformation process to create quality products or services effectively.

Major responsibilities of operations management can be classified into four areas: design, schedule, operation, and control. Figure 6.4 shows the major responsibilities of operations management.

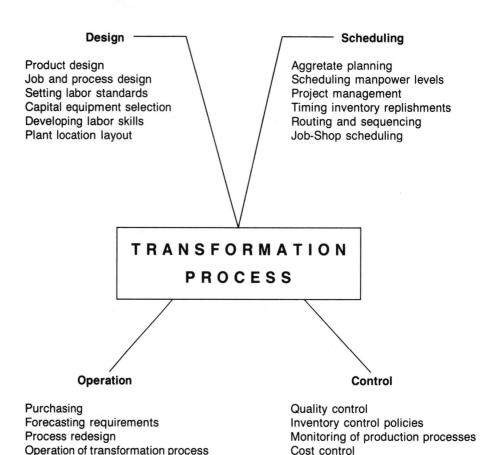

Design ───────┐   ┌─────── **Scheduling**

Product design   Aggretate planning
Job and process design   Scheduling manpower levels
Setting labor standards   Project management
Capital equipment selection   Timing inventory replishments
Developing labor skills   Routing and sequencing
Plant location layout   Job-Shop scheduling

┌─────────────────────────────┐
│   **T R A N S F O R M A T I O N**   │
│   **P R O C E S S**   │
└─────────────────────────────┘

Operation   **Control**

Purchasing   Quality control
Forecasting requirements   Inventory control policies
Process redesign   Monitoring of production processes
Operation of transformation process Cost control
Maintenance   Resource allocation

Figure 6.4 Major Responsibilities of Operations Management

Design

Design is part of the transformation process, which involves equipment selection, type of process, and workflow patterns. Processes of transformation are usually either continuous, intermittent, or "one-shot" projects. Continuous processes are usually highly specialized to produce one type of product or service. Intermittent processes are more general and use a variety of multipurpose equipment. "One-shot" processes, as the name implies, are one-time processes. Other design factors include the layout of the facility and the pattern of workflow needed to achieve a smooth flow of output.

Scheduling

After a process is designed, it must be scheduled to produce the desired product or service at the right time. Scheduling can be long- or short-range planning. Long-range scheduling of forecasts for future demand and devices aggregate plans to estimate needed amounts of manpower, capacity, and materials. Long-range scheduling also includes the tentative scheduling of future activities to meet future demand. Manpower and materials are planned so that they will be available at the right time and the right place to transform the inputs into the desired outputs. This type of scheduling has been discussed in the section on materials requirement planning MRP–II.

Short-range scheduling concerns weekly or daily operations. Sequencing is done on such activities as jobs through machines, vehicles through routes, and patients through a hospital.

Operation

Operating the transformation process is related to scheduling. Scheduling refers to a type of planning that is done before execution; operating refers to the actual implementation of the transformation procedures. Effective operation requires the ability to react to last-minute changes and unexpected delays in planned schedules. In addition, the operating phase involves longer-range decision activities not pertaining to scheduling, such as purchasing, implementing market strategies, and process redesign.

Control

Control involves the monitoring of the transformation process to ensure a quality end product or service. The control mechanism usually consists of some means of measuring the product or service before its completion. Computers are frequently used to monitor inventory levels and can signal an out-of-stock or back-order situation.

Just In Time (JIT)

Just In Time is an approach for performing the production function. It has been used in Japan since the early 1950s. It attempts to minimize the inventory cost by producing smaller quantities. Ideally, a lot size would be only a single item, moving from workstation to workstation. The secret is timing. A supply of raw materials arrives from the vendor "just in time" for the production run; there is no raw materials inventory to speak of. Only small quantities of raw materials are received in a shipment. A single vendor's trucks may arrive several times a day.

Raw materials are started down the assembly line. When the first worker is finished with the task of the first item, the item is set aside. The next worker picks up the item and begins work. If a worker sees a defect caused by a previous worker, the entire assembly line is shut down until the cause of the defect is identified and corrected. The result is improved quality of the items produced.

When workers are ready for a new item, they signal the previous worker by using a *kanban*. Kanban is a Japanese term for "card" or "visible record." The kanban enables the work to flow rapidly by providing a means for pulling the items through the assembly process.

With JIT there is less material in the work flow, and the amount of work space that is needed is reduced. Companies using JIT find they can use more space for production and less space for inventory. Work areas are neater.

Users of JIT do not place a separate purchase order each time they request a shipment. Instead, they usually place a blanket purchase order to cover a specific period, such as a year. Therefore, partial orders can be placed against the blanket purchase order by a telephone call or a form letter.

The contrast between MRP–II and JIT is that MRP–II emphasizes long-term planning and the use of computers, whereas JIT emphasizes timing and the use of noncomputer signals. An MRP–II system can provide the overall planning and coordination effort, and JIT can fit within that framework. It is too early to decide how successfully MRP–II and JIT will interact.

Capacity Planning

Capacity planning is concerned with determining the labor and equipment resources needed to meet the production schedule.

It would not be feasible and would be counterproductive to develop a master schedule that exceeds plant capacity. Therefore, the master schedule is checked against available plant capacity to make sure that the schedule can be realized. Either the schedule or plant capacity must be adjusted to be brought into balance. Capacity planning has always been of concern in traditional production planning. However, in today's manufacturing environment, capacity planning recognition is growing because of its impact on the ability to achieve the master production schedule.

Plant capacity can be defined as the maximum rate of output that the plant can produce under a given set of operating conditions. The operating conditions refer to the number of shifts, number of days of plant operation per week, employment levels, and whether overtime is included. The capacity for a production plant is usually measured in terms of its output units.

Labor and equipment requirements to meet the master production schedule and long-term future needs of the firm is determined in capacity planning. Capacity planning is usually performed in terms of labor and/or machine hours available.

The master schedule is transformed into material and component requirements by using an MRP–II system. These requirements are then compared with available plant capacity over the planning horizon. If the schedule is incompatible with capacity, adjustments must be made either in the master schedule or in plant capacity.

Capacity adjustments can be accomplished in either the short term or the long term. Short-term capacity planning adjustments include decisions on the following factors:

1. *Employment Levels.* Employment in the plant can be increased or decreased in response to changes in capacity requirements.
2. *Number of Work Shifts.* The number of shifts per week can be increased or decreased.

3. *Labor Overtime Hours or Reduced Workweek.*

4. *Inventory Stockpiling.* This factor can be used to maintain steady employment during temporary slack periods.

5. *Order Backlogs.* Deliveries of product to customers can be delayed during busy periods.

6. *Subcontracting.* Letting of jobs to other shops during busy periods, or taking in extra work during slack periods will make the capacity planning system more efficient.

Planning to meet long-term capacity requirements includes the following types of decisions:

1. Investing in more productive machines or new types of machines to manufacture new products.

2. Construction of a new plant.

3. Purchase of existing plants from other companies.

4. Closing down or selling off existing facilities that will not be needed in the future.

Exercises

Homework Assignment

1. Explain the concept of CAB and how this concept is a support for CIM.

2. What is the marketing mix, and what are the components or subsystems of the mix?

3. What is CIPMS and what are the subsystems of it?

4. Contact a nearby industry and review their order entry and purchasing system. Find out the how, who, and when of the process. Is the process computerized, and how does the process flow?

5. Explain the concept of MRP and the advantage of the concept.

6. What is Manufacturing Resource Planning (MRP–II), and how does it differ from MRP?

7. Explain the significance of operations management as support to the manufacturing environment.

8. List and briefly discuss the major responsibilities of operations management.

9. Explain the JIT concepts.

10. Explain capacity planning and its implication for the manufacturing process.

1. CIM is a system integration technology.
   True   False

2. A CIM view of the company's production resources is as a single system in defining; funding, managing, and coordinating all improvement projects in terms of how they affect the entire system.
   True   False

3. The financial function of an organization is concerned with the money flow.
   True   False

4. The marketing function of an organization determines what products and services should flow from the firm to the customers.
   True   False

5. It has been difficult to use the power of a computerized information system in personal setting, advertising, and promotion.
   True   False

6. Advertising is more of a science than an art.
   True   False

7. Demand-based policy establishes a price compatible with the value that the buyer places on the product or service.
   True   False

8. Pricing models should consider both internal and external influences.
   True   False

9. Computer-Integrated Production Management Systems (CIPMS) have traditionally been called production planning and controls.
   True   False

10. The three steps in a control system for the shop floor are purchase release, order scheduling, and order progress.
    True   False

11. Purchasing is the most computerized process in the order entry and purchasing system.
    True   False

12. An MRP interacts with two systems—production scheduling and capacity requirements planning.
    True   False

13. MRP and MRP–II are the same concept.
    True   False

14. Custom and packaged MRP–II software are the same price.
    True   False

15. The MAPICS is the most popular MRP–II package.

True   False

16. There are MRP–II time-sharing services available for companies that do not want to invest in their own hardware or software.

True   False

17. The integration of business components in the manufacturing environment can be viewed as one of the following:

1. CAB

2. CIM

3. CAM

4. CAD

5. CIPMS

18. The orientation of the company's objectives to satisfy the needs and wants of customers for a profit or service is considered the:

1. Planning concept.

2. Marketing concept.

3. Manufacturing concept.

4. Control concept.

5. None of the above.

19. Strategies that enable resources to be used in marketing the company's products and services are called the:

1. Marketing mix.

2. Marketing concept.

3. Financial mix.

4. All the above.

5. None of the above.

20. The marketing mix includes:

1. Product

2. Promotion

3. Place

4. Price

5. All the above.

21. The method by which a company makes its product available to the customers is considered to be the _____ ingredient in the marketing mix.

1. Product

2. Promotion

3. Place

4. Price

5. None of the above

22. Information should flow in two directions—toward the _____ and toward the _____.

    1. Finances, profits

    2. Consumer, manufacture

    3. Saving, profits

    4. Consumer, manager

    5. None of the above

23. Industries tend to follow _____ approach(es) in pricing.

    1. One

    2. Two

    3. Three

    4. Four

    5. Five

24. A basic approach in pricing policies is:

    1. Cost-based pricing

    2. Demand-based pricing

    3. Mathematical-based pricing

    4. Items 1 and 3 only

    5. All the above

25. Major subsystems of CIPMS include:

    1. Production inventory management

    2. Plant shop scheduling

    3. Shop monitoring and control

    4. Plant maintenance

    5. All the above

26. _____ is a system of monitoring the status of production activity in the plant and reporting the status to management so that effective control is established.

    1. Shop floor control

    2. Production control

    3. Scheduling control

    4. Planning control

    5. None of the above

27. _____ is a technique of managing production inventories that takes into account the specific timing of material requirements.
    1. CIPMS
    2. CAM
    3. CAD
    4. MRP
    5. None of the above

28. The evolution of MRP–II includes:
    1. Improved ordering methods
    2. Priority planning
    3. Closed-loop MRP
    4. All the above
    5. None of the above

29. _____ is the effective use of people, products, and processes in the manufacturing industry.
    1. Control
    2. Forecasting
    3. Operations management
    4. All the above
    5. None of the above

30. _____ minimize(s) the inventory cost by producing smaller quantities.
    1. JIT
    2. Demand base
    3. Cost base
    4. Items 1 and 2 only.
    5. Items 2 and 3 only.

ANSWERS TO SELF-STUDY TEST

1.	False	7.	True	13.	False	19.	1	25.	5
2.	True	8.	True	14.	False	20.	5	26.	1
3.	True	9.	True	15.	True	21.	3	27.	4
4.	True	10.	False	16.	1	22.	2	28.	4
5.	True	11.	False	17.	1	23.	2	29.	3
6.	False	12.	True	18.	2	24.	4	30.	5

References

[1] Groover, M., and Zimmers, E., *CAD/CAM: Computer-Aided Design and Manufacturing.* Prentice-Hall, Englewood Cliffs, NJ, 1984.

[2] Hodge, B., *Management-Information Systems.* Reston, Reston, VA, 1984.

[3] Berry, G., "Shop Floor Information System Design and Implementation." *Conference Proceedings,* Autofact 4 Conference, Dearborn, MI, 1982.

[4] Raffish, N., "Let's Help Floor Control." *Production and Inventory Management Review and APICS News,* American Production and Inventory Control Society, July 1981, pp. 17–19.

[5] Schaffer, G. H., "Implementing CIM." *American Machinist,* August 1981, pp. 151–174.

[6] Anderson, J. C., and Schroeder, R. C., "Getting Results from Your MRP System." *Business Horizons,* May–June 1984, p. 58.

[7] McLeod, R., *Management Information Systems.* Chicago: Science Research Associates, 1986.

[8] Piercy, N., *The Management Implications of New Information Technology.* Nichols, 1984.

[9] McFarlan, F. W., and McKenney, J., *Corporate Information Systems Management— The Issues Facing Senior Executives.* Irwin, Homewood, IL, 1983.

[10] Cook, T., and Russell, R., *Contemporary Operations Management,* 2nd ed. Prentice-Hall, Englewood Cliffs, NJ, 1984.

[11] Manji, James, "Handling Systems Move Toward Full Integration." *Automation,* December 1987, pp. 36–40.

[12] "How CIM Keeps One Company Competitive." *Modern Materials Handling,* July 1987, pp. 87–90.

[13] "MRP–II Managing a Manufacturing Company." *Industry Week,* March 23, 1987, pp. 44–46.

PART 2

Manufacturing Systems Integration Techniques and Strategies

The second part of this book covers computer-integrated manufacturing (CIM). It is an extended discussion of Chapter 1. Part II consists of four chapters. Part I discussed the building blocks of CIM. Part II addresses the role of computers in the integration of all phases of product inception; research and design; production planning, controls, and operations; installation; and maintenance of the total manufacturing enterprise, including product support.

Chapter 7 explores the benefits of integrating business functions into CAD/CAM systems to form a CIM system. This chapter also discusses computer integration of business, engineering, and production systems into the manufacturing process over the entire manufacturing cycle.

Chapter 8 addresses important considerations in preparing for CIM. It presents important tasks that aid in moving the manufacturing enterprise toward its goal—CIM.

Chapter 9 provides an overview of the development phases of a CIM system. It also discusses the integration of human resources when an enterprise is developing and implementing CIM systems or any other computer support system.

Chapter 10 discusses the components that should be involved in implementing a CIM system from a design model that has been selected for developing the system.

7

Computer-Integrated Manufacturing

A general overview of computer-integrated manufacturing (CIM) was presented in Chapter 1 where it was defined as a global computerized manufacturing information system. This system provides the conceptual basis for integrating applications, operations, and information flow throughout the entire manufacturing enterprise by means of a common data link. It manages and moves data over a complete manufacturing cycle, as illustrated in Figure 7.1. The manufacturing cycle was defined in Chapter 1. An immediate ability of CIM is its power to integrate data and information about the interrelated activities shown in Figure 7.1 and to enhance parallel and simultaneous model processing. A CIM system is the backbone of a totally integrated enterprise.

The thrust of CIM is the pursuit of higher efficiencies with shorter production cycles all working collectively as a business unit across the entire manufacturing cycle. By successfully linking and integrating all operations through a common database, CIM maximizes and streamlines the entire manufacturing process [1, 68]. A successfully implemented and managed CIM enterprise provides timely, accurate, reliable, and usable information to all functions when needed. The centralization of this information in an integrated database creates an extremely powerful and valuable resource for the entire company.

In addition to bringing manufacturing product processes and operations into focus in Chapter 1, a global CIM definition was introduced for use in this book. Also, the historical development of various CIM foundation technologies was discussed in Chapter 1. Conceptual techniques for computer integration techniques were presented in Chapters 2, 3, 4, 5, and 6.

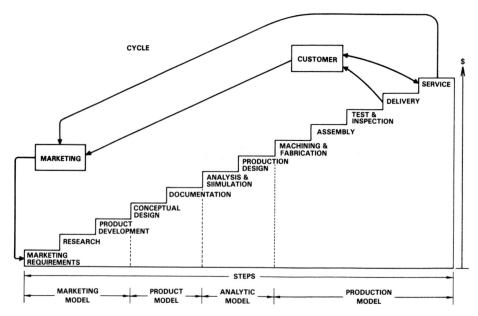

Figure 7.1 A Product Process

Computer Integration of Major Functions

Traditionally, a manufacturing enterprise has been divided into four major functions (business systems, engineering systems, production systems, and human resources systems). These systems have been further fragmented into many divisional and departmental systems (see Chapter 1.) Divisional and departmental activities are set up to obtain efficient and maximum performance. Organization structures vary from enterprise to enterprise. However, good organization should tend to reduce the number of management problems that arise, minimize effort involved, reduce organization friction, promote effective team work, and keep operating costs at a minimum.

Many manufacturing systems organized along traditional functional lines are experiencing problems in meeting customers' demands in a timely manner, producing competitively priced products with better quality and reliability, and providing timely and dependable support services. Such inflexible and inefficient traditional systems are in many cases also losing a greater share of the market, with a resulting decrease in sales. As a result, investors' returns on investment (ROI) decreases, which creates additional problems for the enterprise. Typical of such problems are loss of financial support, reduction in equipment acquisition, increase in time for developing a fully automated factory, and a reduced ability to compete in the world marketplace.

As we have discussed in previous chapters, many manufacturing activities and operations in the major functions can be computerized, automated, and linked together along functional lines through the use of computerized databases and communications systems. Yet, the large-scale sharing of factory data by direct computer transfer across major functional boundary lines is not being implemented in many manufacturing enterprises. However, traditional manufacturing, as it is known today, is undergoing rapid changes in its operations. Leading to these changes is the desire to increase productivity and to produce at lower manufacturing costs. It is also recognized that the hourly rate of the factory worker is becoming a factor in the total cost of the product. Typical manufacturing changes are taking place in the area of computer-aided production and control systems (CAPACS). Typical of such CAPACS are the just-in-time (JIT) system, computer-aided design (CAD), computer-aided manufacturing (CAM), computer-aided process planning (CAPP), manufacturing resource planning (MRP–II), and flexible manufacturing system (FMS). Many industries are well on the way in sharing computer data and information between interrelated CAPACS.

A new approach to manufacturing, CIM technology, is a way of integrating CAPACS (CIMs) and gaining important benefits such as the following:

- Improved productivity
- Lower product cost
- Increased product quality
- Reduced inventory
- Reduced accounts receivable
- Improved manufacturing throughput
- Greater market share and increased sales
- Redistribution of personnel skills

Computer integration of CAPACS and CIMS in a total manufacturing enterprise is a step in implementing CIM. That is, CIM may be viewed as the integration of manufacturing subsystems through the use of computers, modern information technology, and database management systems.

There are numerous examples of modern on-line factories with computer-integrated CAPACS and CIMs. However, a great majority of senior managers of manufacturing companies have not embraced and promoted CIM in their companies. Too many executives still view the computer as an enemy, not an ally, and as a cost, not a competitive asset. Their beliefs are a drawback to implementing CIM, which will not succeed without top management's support.

The conceptual CIM system shown in Figure 7.2 is a nontraditional closed-loop manufacturing system. It provides for the sharing of data through direct transfer between business, engineering, human resources, and production subsystems. This global CIM system also provides for linking individual islands of automation of all functions in the enterprise and enhances control over production operations and processes. Thus, a global CIM system's technologies result in the automation of the information flow in a complete manufacturing enterprise—from order entry through all phases up to and including shipment and field support of the product [2]. As a result, the ultimate goal of CIM is to link automated manufactur-

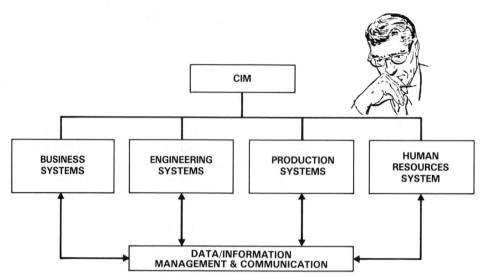

Figure 7.2 Major Functions in a Total CIM System

ing operations such as numerical control, process planning, manufacturing requirements planning (MRP–II), and CAD/CAM through data integration and flow among the many operations. It will then further integrate these operations with the traditional management information systems (MIS) and data processing operations such as financial accounting, inventory control, and payroll. The effective management of these data creates an extremely powerful and valuable resource for the entire enterprise [3].

The implementation of CIM technology often requires that the structure of the manufacturing enterprise be reorganized to efficiently and effectively manage related activities. In most manufacturing companies today, a typical organization chart may look like the one shown in Figure 7.3. Typically, an organization's departments are made up of staff and line management. The organization chart in Figure 7.3 shows a general manager, accounting management, production management, marketing management, and human resources manager, each isolated from the operations of the other managers. Organization charts vary from company to company.

Historically, the host computer was located in the accounting division. As a result, the computer was used to assimilate data for various management reports by using its data compilation capability. Later, operations such as data processing, automated invoicing, and order generation were added to the machine's duties. The host computer continued to be located in accounting because accounting operations were those that were able to derive the most benefit from the computer. During the same time, accounting attempted to build a wall around the computer, making it unavailable to operations in other functions. Such action by accounting led to a multiplicity of minicomputers and microcomputers in other operations because the mainframe computer was not available. Uses by individual operations were also accelerated by hardware and software developments, demands for improved pro-

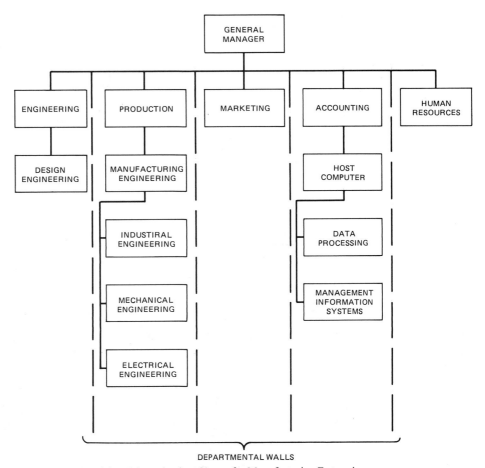

Figure 7.3 A Traditional Organization Chart of a Manufacturing Enterprise

ductivity, and increased help from support groups. Organization of a firm along the traditional lines of the one illustrated in Figure 7.3 is not well suited to manage data processing and factory automation. As a result, many islands of automation were developed.

A CIM system, a nontraditional method of manufacturing, requires another organizational approach to accomplish the following:

- Tie the islands of automation together to achieve CIM's goals.
- Provide a structure for viewing and implementing CIM.
- Tie the organization and automation together via the information system to achieve desired efficiencies.
- Eliminate redundant data and share common data between islands of automation.

A typical organization chart that is suited for a CIM environment is shown in Figure 7.4. Notice the difference between this organization chart and the one in

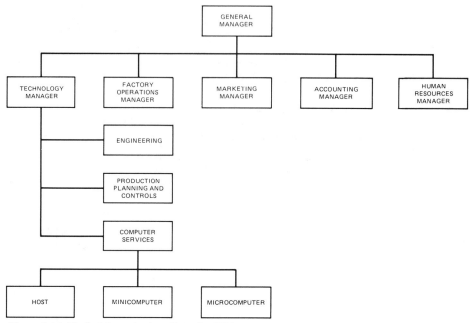

Figure 7.4 A Typical Organization Chart for CIM

Figure 7.3. In Figure 7.4, the production, engineering, and business functions are all housed in one department. This organization chart brings the design and manufacturing engineering people together in the same department with the computer systems people. Such an organization maximizes the use of computer technology in all phases of the manufacturing enterprise.

In summary, CIM has an impact on the organizational structure of an enterprise when the total manufacturing system is integrated. The reorganization structures necessary to put into effect this new manufacturing philosophy improve implementation strategies and management techniques, provide lower implementation costs, result in greater worker participation, and enhance the efficiency of the total system.

A Conceptual CIM System Model

A conceptual CIM model offers more to manufacturing than simply speed and repetition [5, 21]. It also offers better factory control, improved product quality, product flexibility, and increased profits. This nontraditional manufacturing technology represents an enormous opportunity for the manufacturing community to improve its productivity and maintain a competitive position in world markets. This improvement, however, does not come from simply throwing large amounts of money into the manufacturing automation process and saying that the company is therefore integrated. It comes from a well-thought-out plan (See Chapter 8), a logical approach to developing and implementing the plan, and total support from top executives. Implementation requires total cooperation between all functions of the enterprise—that is, team work!!

Authorities seem to agree that an effective CIM system demands changes from the traditional way of running a manufacturing business. They come through the willingness of top management, along with resources from throughout the organization, to aid in making the changes. The complete organization at all levels must be willing to tolerate change.

To develop and operate a global CIM system should be a long-term goal with short-term objectives supported by management and with careful planning and implementation strategies. As a start, a conceptual CIM system is shown in Figure 7.5. It represents a paperless information system. Such a system supports the complete range of business, engineering, human resources, and production activities and operations from strategic planning and customer orders to the delivery of the product, including product support services. The hub of the baseline model uses a common database as the central node. This node is the repository for the data released by

Figure 7.5 An Elementary Conceptual CIM System

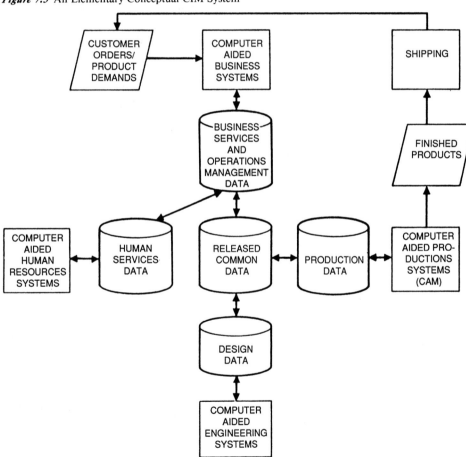

engineering. Data from this node are distributed (copied) to major functional systems to form secondary nodes. The functional users then add values to the data at these nodes to meet their own requirements. Additionally, each function may distribute data to other subsystems for other applications.

This fully conceptual CIM system uses several computers. It is often necessary for a computer in a CIM system to talk to and receive and send data to other computers in the system. Such communication often results in a data translation process and data communications networks. A local area network (LAN) is generally used to pass data from a local database (secondary node) in the common database. The local LAN may be connected to the various computers and their LAN at all levels of the factory to form a common database. To assist in managing the huge common database, intelligent database management systems (DBMS) are used. Thus, an intelligent DBMS and an integrated communications network form the hub of a CIM system.

The illustrated conceptual CIM model should not be viewed as being solely a data processing project. It uses many resources and ties together manufacturing islands of automation and resources such as people, procedures, computers, communication channels, and machines. Thus, it brings together and links elements needed to close the loop on a comprehensive manufacturing information system, including all actions that affect the automated tools on the factory floor, product design, planning, and controls. The development of such a CIM system should be the result of joint efforts of the business, engineering, human resources, and production functions. Typical examples of extended CIM models are shown in Figures 1.20 and 7.6. These models illustrate distributed databases for the subsystem. A difference between the two models is the way the human resource function is used. Figure 1.20 assumes that technology will develop to the degree that people in general will be eliminated. The baseline model in Figure 7.6 assumes that the human element will be around for some time. Both systems are practical and workable and stem from well-developed and well-implemented CIM plans.

Even though the baseline model uses a central node as the common database with several distributed databases, it can be used to develop several various working models. That is, several configurations can be developed around the philosophy of the computer-integrated manufacturing enterprise (CIME). Regardless of the CIM configuration developed, however, typical considerations that should be included are the following:

- User access to distributed data in the network through controlled interfaces.
- A high-speed intercomputer with computer-to-computer communication capability linking the nodes of the system.
- A common graphics language interface to the computer's graphics system.
- Configuration identification and change control throughout the design and manufacturing of the product line involving as-designed, as-planned, and as-built data.
- Automatic generation and distribution of updating transactions to distributed database elements.

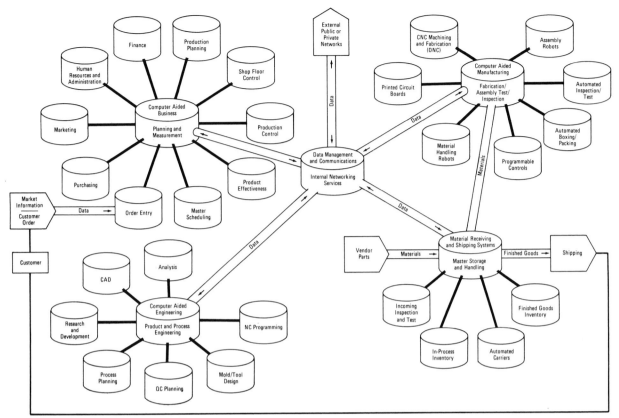

Figure 7.6 An Extended Conceptual CIM Model

- Automatic closed-loop flow of information from design organizations to production and back.
- A good data management and communication system.
- Business subsystems well integrated into the total system.

An immediate advantage of CIM is its ability to integrate such interrelated short-term functions as sales, design, materials management and handling, production, and quality assurance while providing valuable data for cost accounting and for the management information system (MIS) database [6, 4].

Typical examples of the flow of product information (processed data) in a CIM environment are now discussed. As shown in Figure 7.7, all phases of the manufacturing process become computer integrated. The product planning process includes all the early planning activities discussed in Chapter 1. By using time-sharing information, many operations can be performed concurrently as well as serially. As a result, the location of everything in the enterprise is traceable at the touch of a few computer keys.

A CIM system interacts with such top-management functions as financial planning, marketing, and such heretofore routine functions as order entry, maintenance, shipping, and receiving. Also, CIM has the ability to permit the user to respond quickly to changes in demand and marketing conditions, thus reducing warehouse inventory when demands are low and a loss of sales when demands increase.

The computer can take a message of an external order entry station and direct the appropriate machines and supporting elements to make the desired quantity of the part. It can download CAD data to prepare CAM routines and checks to be sure the part is made as directed. It can also call up a bill of materials processor (BOMP), direct assembly of components, order their replacement, and compile the sequence of machine steps necessary to produce the parts. Sensors report on quantities, measure tolerances and report them for statistical process control (SPC), and trace the process so that managers can evaluate it for quality or productivity.

A typical CIM working model developed from the conceptual model shown in Figure 7.6 is illustrated in Figure 7.8, which shows the scope of CIM that we have discussed. The factory management subsystem (FMS), a CIMS, is the centerpiece of the global CIM system. This FMS module manages all the manufacturing resources shown in the figure in order to produce finished products. It is a highly sophisticated module suited for a total integrated automated manufacturing enterprise.

Figure 7.7 A Simplified CIM Cycle

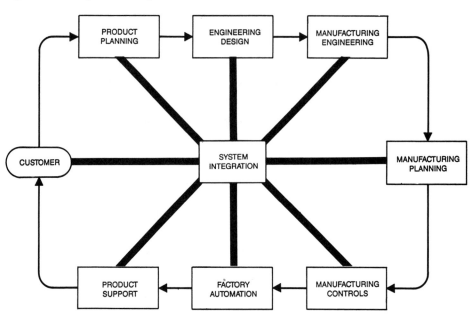

Functional Organization

Developing and implementing a CIM system as illustrated in Figure 7.5 is a tremendous task. A company can accept the concept of CIM without realizing the disciplines that will be required to achieve a CIM objective. It should be remembered, however, that CIM is not a data processing project alone. It involves the use of many resources, and such use must be supported by decisions and actions by top management.

Experience has shown a direct relationship between the chief executive officer's (CEO) level of involvement and the eventual success or failure of an implementation plan. Consequently, the involvement of management at all levels of the enterprise up to and including the CEO, the greater the chance for success. Therefore, it is advantageous to have the highest possible decisionmaking management directing the project. Usually, lower-level management will abide by upper-level management decisions.

From our discussions so far, it is apparent that the organization structure of an enterprise has a vast impact on CIM development (Chapters 8 and 9) and implementation (Chapter 10) processes. As a result, changing from a traditional organization structure to a nontraditional structure must involve top management and everyone below. A nontraditional organization chart should look similar to the one shown in Figure 7.4. Notice that this organization chart shows the highest executive

Figure 7.8 Computer-Integrated Manufacturing

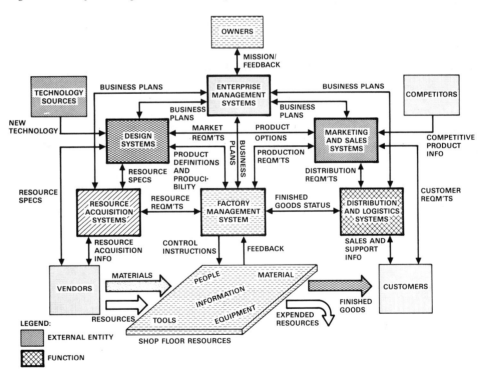

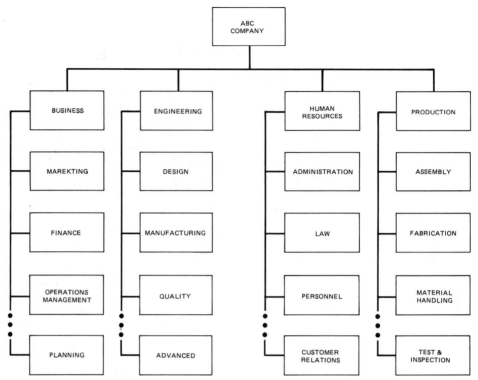

Figure 7.9 An Operational Structure in a Traditional Manufacturing Organization

officer as being in charge of the major functions involved in CIM activities and operations. Although this is a typical functional diagram, it is by no means universally applicable. Each manufacturing enterprise must develop its own version of a functional diagram to carry out its strategic plans and meet its production requirements.

An enterprise organized in the traditional way (Figure 7.9) should not have to make drastic changes. One approach would be to coordinate the existing CAM, CAB, CAE, and computer activities under a high-ranking executive officer, such as a senior vice president. This approach to organizing CIM functionally is illustrated in Figure 7.10. The executive officer would be in charge of the CIM system for all functions. A vice president would be assigned to each function. Without such an alignment of major functions, more islands of automation might be developed and implemented. The implementation process of a CIM system derives its success from top management commitments and involvements at every point, place, and time.

Some advantages of the organization alignment shown in Figure 7.10 are as follows:

- The familiarity of all knowledgeable workers with the business design and manufacturing processes is used.
- Communication between functions is increased.

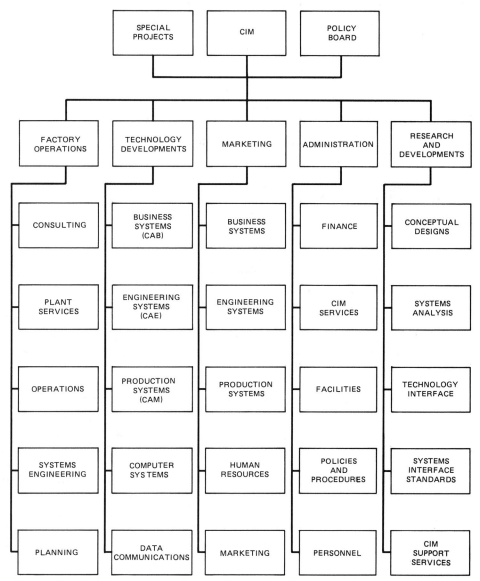

Figure 7.10 A Typical Organization Chart for Operational CIM

- With increased cooperation, there is less friction between functions. More workers cooperate and work toward a common cause.
- The use of computer technology in all phases of manufacturing is maximized.
- Duplication of manufacturing efforts is reduced.
- The use of data processing, communications protocols, and database standards is increased.

Computer-Integrated Manufacturing

- The duplication of equipment and accessories and the cost of initial investment are decreased.
- The company can move toward a computer-integrated manufacturing environment with less difficulty.
- A high-level executive officer coordinates the developing and implementing of systems to bring about computer-integrated manufacturing.
- Manufacturing computer specialists and manufacturing specialists are brought together under one high-level executive officer.

Computer Hierarchy Controls

The technologies of CIM provide for integrating existing islands of automation in a manufacturing enterprise under one system and reduce the creation of new islands of automation. This integrated global system consists of advanced manufacturing technologies and support systems that use computers as an integral part of their control (Chapter 2). Computers interconnect all the subsystems (CIMs) of all the major functions, as shown in Figure 7.6. They also control standalone manufacturing systems such as robots, machine tools, materials handling equipment, test and inspection, laser-beam cutters, and data collection in a real-time mode. The integration of operations of standalone manufacturing systems is referred to as a flexible manufacturing system (FMS). Please note that FMS carries two different meanings in this text (see also Chapter 5). Along with business and engineering systems, all NC machines, automated workstations, materials handling and transfer systems, and data collection systems are controlled by computers (see Chapter 4). As a result, controlling the many independent systems requires a very complex control system.

The architecture of a typical CIM computer control system is illustrated in Figure 7.11. This system shows the relationship of the major functional computer to subsystem computers, which are selected and arranged so they functionally connect the many islands of automation into a cohesive operation. This architecture permits a flexible CIM system to be designed to meet high manufacturing standards and requirements. A CIM system requires close control over the complex systems that consist of an integrated assembly of production and control subsystems.

To meet the complex system's control requirements with communications capabilities, a distributed control technique for manufacturing operations and processes is used in the CIM control architecture shown in Figure 7.11. Such a control technique (Figure 7.12) is one of the most demanding characteristics of a CIM system. It takes into account and includes all control actions that affect the automated tool on the factory floor by linking and integrating all operations through a common database.

Four major levels of computer controls form the architecture of this hierarchy. In the illustrated concept, the first level of computer control (Level 0 in the hierarchical structure) is the corporate computer. A mainframe computer provides corporate management with information necessary to formulate and implement the

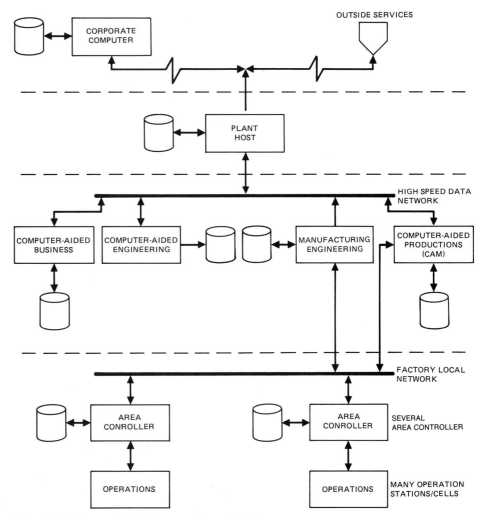

Figure 7.11 Illustrated CIM Computer Control Architecture

company's business plans. This is the highest level of computer control in a CIM system. Data are collected from the various plant sites by this corporate-level computer. The data are then analyzed, synthesized, compiled, and made available to the management in simplified usable reports, tables, graphs, and so on for strategic planning and the implementation of plans. The corporate computer may also be shared with corporate-level and auxiliary planning and support groups.

The second level of computer control, Level 1 in the hierarchical structure, is the central plant host computer. A plant host computer is generally located in each plant in the total CIM system and is generally a large-scale data management system based on a mainframe. It provides on-line storage capabilities needed to store and manipulate the vast amounts of data that are required in a CIM environment. It

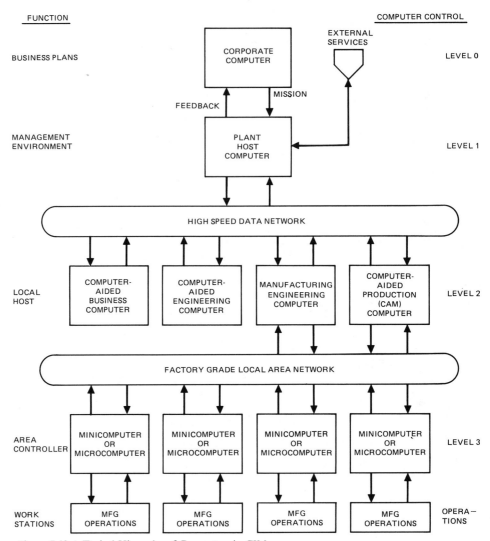

Figure 7.12 A Typical Hierarchy of Computers in CIM

must also provide the capability to store large amounts of manufacturing data such as drawings, process plans, robot and numerical control programs, tooling and fixture information, three-dimensional part modeling, and so on. The plant host computer shares data with the users of the plant's major functional computers. That is, each functional computer can be accessed by the host computer.

The plant host computer must also perform such functions as tracking when data are released from engineering to production, transmitting data to a variety of output devices such as plotters and printers, collecting data from plant operations, and summarizing operations and preparing timely reports for plant management.

The third level of computer control, Level 2, provides services for each major function. These computers are often referred to as satellite computers and are generally minicomputers. At this level, they perform the familiar automated engineering, business, and productions operations. This is where part design, analysis, graphic programming for numerical control and robotics, and production planning and control will continue to take place.

The purpose of Level 2 computers is to serve in a supervisory capacity for the functional operations in the plant. This capacity includes coordinating activities of smaller computers under the command of the Level 2 computers. Data are collected on the individual machine tools, test and inspection stations, materials handling equipment, and work centers. The computer then sends instructions back to the separate processes and workstations for action.

The fourth level of computer control, Level 3, is on the factory floor in close proximity to the processes it monitors and controls. In many cases, the computer is an integral component of the production machine tool. A typical example is a computer numerical control (CNC) machine. Computers and controllers also communicate with Level 2 in the computer hierarchy.

Desktop engineering workstations are also found at Level 3. They provide low-cost access to the valuable central information resource at Level 2. Such workstations enable engineers, designers, quality control personnel, and other design and manufacturing workers simultaneously to expand the central information resource that is a vital key to CIM. This access allows a greater number of individuals to perform a wide range of functions throughout the company—from the design office to the factory floor.

Advantages of Levels of Computer Hierarchy

Several advantages of hierarchical computer control structures are the following:

- Process monitoring data, production status information, and operational data are passed upward through each level of the computer system, and at each step, bulk information is filtered and integrated for more efficient use.
- Commands and schedules are passed down from the corporate computer throughout the rest of the processing structure.
- Excellent shop-floor monitoring and controls are provided.
- Added value comes from the productivity benefits gained at each step of operations as a result of sharing and capitalizing on information from the same centralized data resource.
- Product information is more accurate, so product quality and reliability are enhanced.
- Impounds communication among design, production, business systems, and other major groups of the company.
- There is greater predictability in every phase of manufacturing because extensive information on each operation is stored and readily accessible.

Database Management Overview

A major thrust of CIM is linking and integrating all manufacturing operations through a common data link (see Chapter 2). Using this process, CIM maximizes and streamlines the entire manufacturing process [7]. Data are transferred from one node to another node when requested for processing by a basic function. To maintain data integrity, the data must be managed throughout the database system.

A database management system (DBMS) (see Chapter 5) is a software product that controls a data structure containing interrelated data stored so as to optimize accessibility, to control redundancy, and to provide or offer multiple views of the data to multiple applications programs. A DBMS also implements data in dependence to varying degrees. Utilities, multiprogramming capabilities, and the sophistication of the data dictionary vary with different products and different vendors.

A DBMS is a software package designed to operate interactively on a collection of computer-stored files—that is, a database. Its primary operation is to count database records that have user-specified common characteristics and retrieve those records for further processing and display.

A CIM database system is organized with a complex set of software. A typical DBMS includes a high-level compiler-like language that can be used to describe the location, contents, relationships, and security level of data that are stored in the database. It also has subroutines that are used to manipulate and extract data from the database, such as "Query," or some other English-like language that acts as an impromptu report generator. Off-line routines, or utility programs, are used to load the database onto disk from other media and to duplicate the database as a backup.

A DBMS is the heart of the database system. It supplies each functional activity with the requested data and provides a common data storage and retrieval structure (see Figure 7.13, which shows that a DBMS manages data to and from each node.)

DBMS Architecture

The architecture of a DBMS is structured and presented to end users in such a manner that they can view needed data in a logical sense (Chapter 2). That is, the users need not be concerned about the technical aspects and many other laborious tasks connected with the architecture. Rather, the users view the data in the context of their own requirements.

Three database models commonly used in a CIM environment (Chapter 5) are [8,22] the following:

- Plex or network
- Relational
- Hierarchical

The means by which data records are associated is the essential distinction among the three distinct types of databases commonly recognized. The organizational dif-

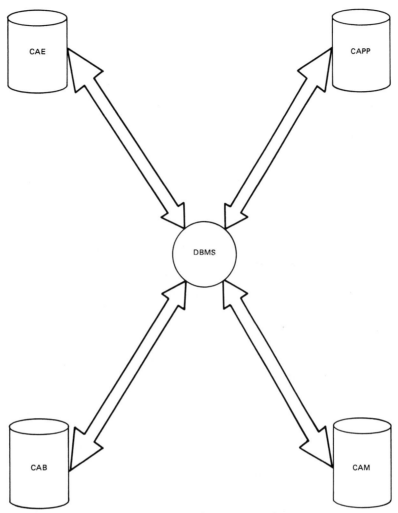

Figure 7.13 DBMS Techniques Integrate Manufacturing Data

ferences are substantial, because the structure of the databases is oriented toward different objectives. Each, however, provides the essential virtues of formal database technology. The relational database model has gained considerable popularity and is used in CIM because of the simplicity of its user views. In its simplest form, the relational database is presented as an organized number of two-dimensional tables (also called flat files). As illustrated in Figure 7.14, a relational database is comprised entirely of tables of data, with each table resident in a system file. Rows of the tables are records (or "tuples") of several fields of entries each.

Each table is called a "relation" because the mere presence of items in a record indicates that they are related by the relationship represented by the table. Thus, for example, a table might be named "Inventory of Parts" and contain only columns of component part number, description, location, number in stock, price, and stock

Computer-Integrated Manufacturing **245**

	COMPONENT PART NO.	DESCRIPTION	LOCATION	NO. IN STOCK	PRICE	STOCK DELETION DATE
TUPLES →	634	* * *	NY	50	10	7/82
• →	788	* * *	LA	200	20	8/82
• →	996	* * *	SF	100	15	10/82
• →	1500	* * *	KC	350	5	1/83
• →	1634	* * *	LA	600	10	7/83

Figure 7.14 A Relational Database Model

deletion data. All items in the same row are related to the component part number. Deleting the tuple simply indicates that the component part number is no longer in stock (or "deleted date" could be printed under stock deletion date), and it does not disturb any other data about the component part.

The information in a database is stored, retrieved, and operated on by capabilities in the DBMS. It is through these capabilities that terminal users or using programs can access the data. Regardless of the database architecture, the key is to use the same product data in all operations. That is, value can be added to the product data by users of the data. As a result, this approach requires discipline and commitment to the *shared database* concept as shown in Figure 7.15. This concept allows the various functions to have access to the data and, at the same time, provides a degree of insulation. This insulation of the data from the users and programs maintains a high degree of independence and portability, allowing systems to take advantage of technological advances that have alluring software.

Distributed Databases

Distributed database concepts (Chapter 5) may be viewed as an arrangement of separate databases dispersed to various areas within the manufacturing enterprise. These databases form a complex database system in which all are working in a cooperative manner. This database system is used in a CIM environment rather than the conventional single database at a single location. Examples of distributed database concepts are shown in Figures 1.20 and 7.6. Such an arrangement of data-

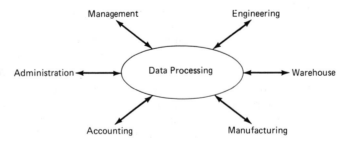

Figure 7.15 Shared Database Concept

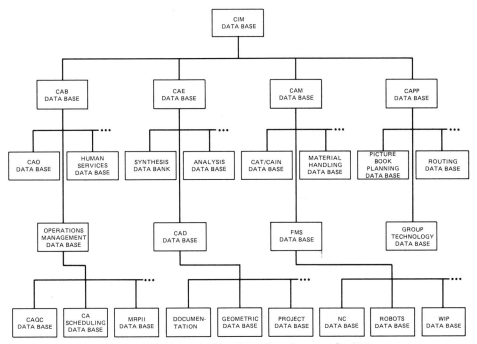

Figure 7.16 A Conceptual Hierarchical Distributed Database Structure for CIM

bases lessens the load on a central database system, increases its versatility, and increases the physical access to the data.

Distributed database management and communications represents one of the more interesting and challenging problems of CIM. Considerable care must be taken in implementing distributed databases if the potential benefits on a distributed system are to be realized. It should be remembered that a database system is the communication element linking each major node in the system and making data circulate from one node to another node. Thus, the aim of a CIM database system is to support major function modules and provide the necessary interface among functions and users.

Refer again to Figure 7.8, which shows various CIM activity modules and their relationship to one another. In this environment, each function is computerized at some level with its own database supporting the assigned operations and working cooperatively with other databases to form a total integrated system.

Hierarchically distributed database systems are becoming a commonplace in CIM environments. A typical hierarchically distributed database system is shown in Figure 7.16. Such a system shares data with and communicates with such major nodes as computer-aided business (CAB), computer-aided engineering (CAE), computer-aided process planning (CAPP), and CAM. These data are further distributed to other dedicated computers for specialized processing.

Several types of data distribution techniques can be implemented in a CIM system. They are of two general categories: partitioned and replicated. A partitioned database technique is illustrated in Figure 7.17. In this scheme, the data are split into

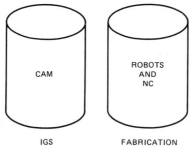

Figure 7.17 A Partitioned Database Technique

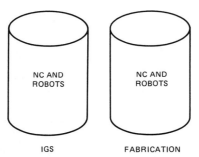

Figure 7.18 A Replicated Database Technique

pieces, or partitions, and assigned to selected sites in the network. The partitioned databases may all be of the same data structure, format, or access method, in which case the system is a homogenous partitioned database. Otherwise, it is a heterogenous partitioned system if the data consist of different structures, formats, or access methods.

A replicated database technique (shown in Figure 7.18) stores multiple copies of the same data at different sites in the network. If the structures among the replicated data are the same, they are called homogeneous replicated data. In many cases, the duplicated data are reformatted or placed under a different access method to fit local needs. In this case, they are called a heterogeneous replicated system.

Databases may also be distributed vertically or horizontally. The vertical topology usually appears as illustrated in Figure 7.19. In this arrangement, data are distributed in a hierarchical manner. Detailed data may be stored locally and generally; aggregated data or less detailed data are stored farther up in the hierarchy. A horizontal topology system is shown in Figure 7.20. This system is a peer-oriented network wherein data are distributed across peer sites in the network.

Summary

A distributed database is one that is spread throughout the computer network. Each data item in these systems is ordinarily stored at its most frequently used location, but it remains accessible to other network users. Distributed systems provide the control and economy of local processing with the advantages of information

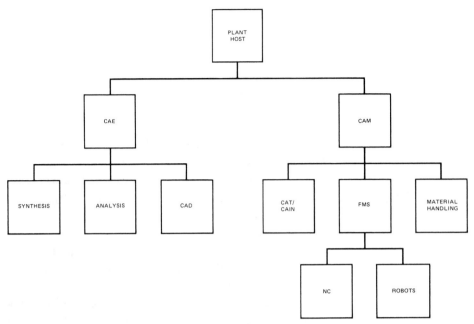

Figure 7.19 A Vertical Topology Database Arrangement

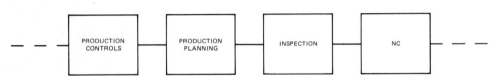

Figure 7.20 A Horizontal Topology Database Arrangement

accessibility over a geographically dispersed organization. These systems can be costly to implement and operate, however, and they can suffer from increased vulnerability to security violations.

The hub of a CIM database system should be a well-developed CAD/CAM database. The database system of a manufacturing business should be well integrated into the CAD/CAM database, forming a well-integrated database. The CIM database should be intelligent, having a high degree of flexibility and responding quickly to changes, including changes in product and process technologies. Also, the system should provide a concurrent, cooperative, and communicative environment for all manufacturing activities.

An index to all the items in an organization's database is called a data dictionary. A data dictionary is actually a database containing information about the database. It lists such items as the types of information in the database, which programs use which data, which user groups require which reports, and which entities of each type exist in the system.

A DBMS provides a data structure in which information can be stored and accessed for many applications. For example, all information regarding inventory of parts is contained in a database file marked INVENTORY. A CAE program accesses this file. While performing its own specific function, the inventory control program updates the INVENTORY database file with current information from a purchasing period so that any other programs needing information will have the up-to-date data.

Some reasons for distributing data are the following:

1. Placing data at the local sites can reduce the amount of data transferred among the network computers, resulting in reduced communications costs.
2. Local access of data can improve the response time needed to obtain the data because the delays of remote transmission are eliminated.
3. Distributed databases can give increased reliability to a system because the data are located at more than one site. The failure of a node need not close down all data access in the network.
4. The provision for local storage gives users more control over the data. As stated earlier, increased quality is likely to result from user involvement and in user control of the data.
5. Distributed databases present a challenging technical problem to the user in control of using the distributed database. The owners, using agreed standards, will use various means to make it work and function effectively and efficiently. This technique points up to the fact that local management is in control of the process and database. While this statement is made somewhat tongue in cheek, the author knows of instances when organizations allowed technicians to move to the distributed data environment because the technicians wanted to "make it work." To be sure, this is not a very good reason, and it once again points up to the need for management control of the process.

Data Communications

Communications between factory control systems, equipment, and various levels of management is of paramount importance in CIM (Chapter 5). The major functions of CIM and the islands of automation must be integrated, and communications is the means of support for the six vital information, coordination, and control functions required for an integrated manufacturing hierarchy [10, 3]:

1. Distributed control
2. Monitoring
3. Data acquisition
4. Supervisory control
5. Program support
6. Management information

TABLE 7.1 Levels of Factory Communication Networks

Level	Hierarchical Position	Description/Application
0	Highest	Corporate, interfacility (wide area network, fully integrated), business plans, low speed, modem-type connector.
1	Center A	Plant, interfacility (plant), (bridges CAM, CAM, CAPP, CAI, environment, MIS, high speed data LAN, high message integrity.
2	Center B	Interdepartmental, local host (Links specific plant cells/operations) Factory grade LAN, wiring suitable to the plant floor, insensitivity to electronic noise.
4	Lowest	Factory floor, departmental/cell level, LAN application-dependent, low cost.

Each of these functions has a different set of requirements, which makes it very difficult for a single communication network to meet all requirements. Additional requirements are placed on the data communications system by having machine toolers talking to process controllers, programmable controllers, robots, materials handling systems, and so on—thus, overtaxing the system.

The factory communications system also provides a communications link between machine operators and the host computer through shop floor terminals. This system provides communications between the operators and various support functions, such as supervision, methods, production control, quality control, and maintenance. It also monitors shop status, keeping track of materials, orders, and personnel.

Because the amount of data and the need for it to be timely vary widely within the mix of the many CIM activities, the system must have a number of kinds of networks to provide complementary solutions to integrate these functions across the plant hierarchy. Also, today's technology is such that no single network will solve all the factory's communication problems.

A solution to the factory's communication system consisting of several levels of communication networks was shown earlier in Figure 7.12. Each level should be designed to meet the requirements of the network components for the network to function effectively and efficiently. The network functions are given in Table 7.1.

Data Communication Standards

Data communication standards should be used when possible in network designs. Data communication standards are just as important in the factory as in the office. Obviously, pieces of office equipment must be able to communicate with one another, but a lack of standards in the factory environment is an almost insurmountable obstacle to automation. If there is a lack of factory standards, the many little islands of automation, composed of proprietary systems cannot, by their very nature communicate or interact easily with systems from other vendors. Users are forced to buy piecemeal solutions for the individual protocols to make the factory function efficiently. The solution is to develop communication standards through factory automation.

Vendors and factory users can no longer wait for the standards committees to come up with a solution to local area network protocols and hardware and software interfaces. They must work together and with standards committees to solve the problems. Critical standards to be determined involve network management—that is, error correction, the movement of information to devices, how the devices will get that information, who talks first, and who has control of the session.

A versatile and widely supported network with common sets of rules is essential in CIM. The manufacturing automation protocol (MAP) and technical office protocol (TOP) share a set of common communication rules to form a communication network (MAP/TOP) well suited for CIM. The MAP/TOP broadband network can link the office management, factory center, system, and cell levels of control. Hierarchical control should be implemented in CIM. However, the peer-to-peer communication protocol in MAP/TOP can be used to provide the flexibility needed for interequipment communication. Establishing communications between various kinds of equipment is a difficult task.

Manufacturing Automation Protocol

For factory-floor operations, MAP is growing in popularity. It is viewed as a standard for the shop floor network. The objective of MAP is to remove factory automation roadblocks by implementing a communications standard in various manufacturers' hardware processors and plant floor devices. Its purpose is to speed the evaluation of nonproprietary integrated communications networks to support multivendor manufacturing automation systems based on national and international standards. Such a network as MAP will link all computers and automated devices in a plant and thus achieve integrated automation.

The Production Process

The manufacturing activities and operations in the manufacturing cycle (see Chapter 1) evolves into an automated closed-loop CIM system from product innovation to product delivery and service. The cycle begins with the customer (Figure 7.5), who defines the requirements. This is a marketing operation as shown in Figure 7.8.

The definition of requirements resulted from the input of many variables such as customer demands and requirements, competitive product information, the enterprise's business plans, and advances in technology. Through an interface device to the marketing and sales systems module or through manual entry, a production order is launched into production. Generally, this module provides control over such operations as entry of order, changes to orders, scheduling changes, and setting priorities to accomplish the current production plan.

Preliminary product requirements from the marketing and sales module are sent to the design systems module for product design. Using computerized iteration processes with such modules as marketing and sales systems, factory management systems, enterprise management systems, vendors, and resource acquisition, a product model is developed and refined by the design systems module. Figure 1.6 shows a view of the product processes after the preliminary product requirements have been identified. The refinement of the definition occurs throughout the design and engineering phase before the product is released to production.

The product model is stored in the common database, generally an intelligent database. As a result, an authorized user may take a close look at three critical elements in the automation effort, which are [7, 17]:

1. Identification of KEY DATA in each step of the cycle:
   - Textual (dimensions, bill of material, etc.)
   - Graphic (geometry, shapes, and surfaces)
2. CROSSOVER POINTS for integration during the process:
   - Marketing to development
   - Development to engineering
   - Engineering to production
   - Production to marketing
3. The importance of model integrity:
   - Control of key variables

Once the design is completed, the product model is released to the central database in production. This distributed database becomes a value-added database because production information is added to the product model. Such data as process plans (CAPP), CAD/CAM graphics, and classification and coding are added. Some advantages of adding values to the data at this time are the following:

1. Enhanced description of the sequence of operation steps.
2. Lower production costs.
3. Higher use of expensive production equipment and planning time.
4. Pictorial process planning sheets.
5. Enhanced process plans.

The feasibility of producing the product economically is verified by manufacturing engineering. In some cases, the ability to manufacture a part may be verified during design and engineering.

Tooling to fabricate, assemble, and test and inspect components and finished product are selected. Test and inspection techniques are selected by using computer-aided testing (CAT) and computer-aided inspection (CAIN) systems. During these operations, the computer searches for various design models that can be modified easily to meet the production requirements. The performance of proposed tool designs can also be simulated. Again, it is possible to perform simulation at the proposal stage as well.

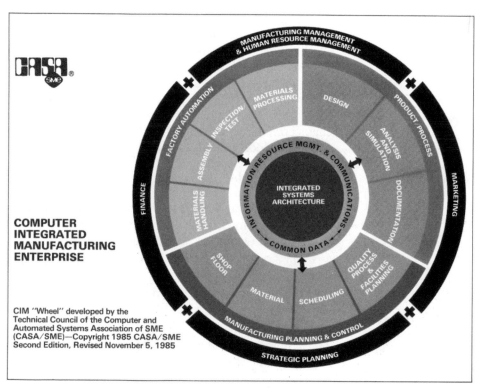

Figure 7.21 The CASA Wheel of SME. *Reprinted courtesy of the Society of Manufacturing Engineers.*

For those parts of the system that will be built on a numerical control (NC) system, control information is automatically produced by the computer. The NC tool paths can be verified on the cathode ray tube (CRT) terminal, thus reducing the cost of NC part programming. From the same design data, operations of robots can be observed and their performance can also be simulated on the CRT.

The FMS schedules all manufacturing resources and manages these resources to produce finished products. Factory floor operations are performed by automated systems such as flexible manufacturing, data collections, robotics, CAT, and CAIN.

Interface between the FMS and other modules in the CIM system is important. The FMS is the centerpiece of CIM and has iteration capability, under computer program, with the other five modules of the system.

Working with other systems such as design, enterprise management, factory management and vendors, the resource acquisition systems module acquires and provides the resources to the factory floor. The control instructions are provided by the FMS.

Two other important modules of a CIM system are the enterprise management system and the distribution and logistics systems. The distribution and logistics management module aids in providing field support, supplies, and equipment. An example is the rapid provision of spare parts—which could reduce standing inventory requirements. The enterprise management system module provides management business plans to all modules in the system. Collectively, these models and factory resources, along with certain external entities, form a computerized automated closed-loop system. A careful review of this technology will help to ensure economic survival in an increasingly competitive international marketplace.

In summary, CIM is a global concept that includes far more than the traditional manufacturing processes found in the factory. It integrates all data processing functions within the company, including financial accounting, purchasing, inventory, distribution, payroll, engineering, and management, as well as traditional manufacturing operations. Figure 7.8 shows the interrelationship between CIM's "basic" functions. Notice that the differences between the ICAM wheel shown in Figure 5.2 (the production cycle) and the wheel in Figure 7.21 by the Computer and Automation Systems Association (CASA) of Society of Manufacturing Engineers (SME) for a manufacturing cycle or CIM cycle. Each wheel consists of modular subsystems controlled by computers that are interconnected to form a distributed system that performs the basic functions shown in the wheel. As a result, the wheels may be implemented in modules, starting with the system's integrated architecture.

Benefits of CIM

The integration of CIM technologies into a global CIM system can have a pervasive and profound effect on manufacturing organizations. Through CIM technology, data are integrated and shared by all segments of each major function: business, engineering, and production. This is known as integrated business, engineering, and production (IBEP).

Important Benefits of IBEP Techniques

The IBEP techniques of CIM technology offer many benefits. Typical of them are the following:

- Improved manufacturing planning
- Production flexibility
- Lower production costs
- Increased product quality
- Greater market share for product

These benefits demand effective planning, implementation, and control processes of any and all manufacturing resources of the entire company. A well-planned system comes through a willingness on the part of top management to be involved in all processes. Planning for tomorrow's factories to be a truly computer-integrated manufacturing enterprise (IBEP) should be manufacturing managements' greatest challenge today.

Improved Manufacturing Planning

The use of a common, integrated, and shared database from the inception of a product through delivery and service support to the end user enhances better manufacturing planning. Planning data from every segment of manufacturing during every phase of manufacturing can be closely tracked and stored in an IBEP data system. Thus, data on each planning operation is available for use by all who need it. As a result, greater predictability can be made in every phase of manufacturing because this extensive stored information on each operation is readily accessible. Also, this information gives better control over many manufacturing processes, leading to higher quality products, and more effective use of manufacturing resources.

Production Flexibility

Production flexibility allows for a wide variety of specific product designs, product mixes, and a shorter product life cycle for all phases of the production cycle. A production life cycle includes all activities from a decision to manufacture through product shipment. Production flexibility also provides the production of a wide variety of volumes of product mix at low cost with a high degree of responsiveness. The improved throughput that results from the responsiveness of the system gives a shorter product turnaround time through the optimum use of manufacturing resources, sharing the same product data, and minimizing inventories, machine breakdowns, and work stoppages caused by missing parts or materials.

In summary, production flexibility permits the production of aggregated volumes of high-quality dissimilar parts and product mixes at low costs and at the same time allows for more customized product designs, production, and marketing.

Lower Product Cost

The linking of CAD/CAM systems (Chapter 5) with computer-aided business systems (Chapter 6) increases the total system's productivity. Such integration, IBEP, results in lower product cost per item. This reduction cost comes as a result of sharing and capitalizing on information from the same IBEP data system. Typical examples of lower manufacturing operating costs are in design, planning, machine tool programming, tool and fixture engineering, materials handling, quality assurance, production controls, marketing, and product support.

Increased Product Quality

Computer-aided design enhances quality in the design itself and often provides superior designs through computer analysis, and CAM assures a high level of repeatability, which further improves product quality with the aid of such systems as CAT and CAIN. The fast response time of computer-based systems supports total quality control (TQC) and thus results in increased product quality for customers.

Greater Market Share for Product

Through CIM, a manufacturing enterprise can target marketing to a degree never before possible. The particular needs of various customers can also be met through customization. That is, a product can be custom tailored for smaller market segments and at the same time serve a more fragmented customer base. Another marketing benefit is that faster response to market demands because of the ability to quickly change the company's product mix provides shorter product cycle life, and makes efficient use of manufacturing resources resulting in increased manufacturing output.

In summary, CIM changes the way of doing manufacturing business. It maximizes and streamlines the entire manufacturing process by computers managing information, time, people, money, machines, and materials. As a result, CIM implementation results in improved productivity in the entire manufacturing enterprise.

Exercises

Homework Assignments

1. Explain the effects of CIM on the traditional manufacturing process.

2. List several subsystems, such as an automated quotation subsystem, in a complete CIM system.

3. List and discuss several arguments *against* CIM.

4. List and discuss several arguments *in favor* of CIM.

5. Define IBEP. Discuss its application in a CIM system.

6. List and discuss several CIM implementation strategies.

7. List and discuss several benefits gained from an operational CIM system.

8. Propose a functional organization chart for a manufacturing enterprise that may overcome some of the apparent problems during a CIM implementation process.

9. a. What is an intelligent database system?

   b. Discuss the need for an intelligent database system in a total CIM system.

10. Develop the architecture for CIM showing the relationship between the major manufacturing functions.

11. a. List the control levels in a manufacturing enterprise.

    b. Discuss advantages and disadvantages of using control levels.

12. a. List some problems associated with implementing a global CIM System.

    b. Discuss each.

13. a. Define an intelligent database.

    b. Give some examples of its application.

14. Explain the differences, if any, among CAPAC, CIMS, and CIM.

15. Explain the difference between the two meanings of FMS.

16. List several benefits of CIM.

17. What does distributed systems mean?

18. Describe a typical hierarchical database and relational database.

19. Define (a) MAP, (b) TOP

20. Why are standards important in CIM?

Self-Study Test

1. The "paperless factory" is often cited as the ultimate goal in the world of manufacturing.

    True   False

2. The hub of a baseline CIM system should use a common intelligent database system as the central node.

    True   False

3. With a truly CIM system, some emergency parts can be manufactured and shipped the same day an order is received.

    True   False

4. Typical CAPACS in CIM are:

    1. CAD, CAPP, CAM and MRP–II

    2. Quotation, order entry, cost estimating, and accounting

    3. MRP–II, FMS, CAT, and CAIN

    4. Items 1 and 3.

    5. All these.

5. A CIM system usually requires an organization approach to manufacturing in order to:

    1. Bring the islands of automation together through a common data-link system.

    2. Provide a structure for viewing and implementing CIM.

    3. Eliminate redundant data and share data between islands of automation.

    4. All these.

    5. None of these.

6. Many executives today view the computer in manufacturing as:

    1. An enemy and not an ally.

2. A cost and not a competitive asset.

3. Causing delays in completing projects and not as a tool that aids manufacturing production.

4. All these.

5. None of these.

7. Typical measures of CIM performance include such fundamental(s) as:

1. New product development lead time

2. Cumulative manufacturing lead time

3. Inventory turns

4. Quality levels

5. All these.

8. MRP–II stands for:

1. Material Requirement Planning

2. Material Resource Planning

3. Manufacturing Resource Planning

4. Manufacturing Requirement Planning

9. CAPACS came about as a result of the marriage between _____ and _____.

1. Manufacturing production, production controls

2. Manufacturing technology, computer technology

3. Production controls, manufacturing technology

4. Production controls, computer technology

5. Improved technology, cost effectiveness

10. An information system

1. Consists primarily of technology and equipment

2. Consists of people, rules and procedures, data, hardware, and software

3. Guarantees correct decisions in the future

4. Is concerned fundamentally with processing

11. The correct order of usable elements of the data hierarchy, from the broadest to narrowest, is

1. Database, field, record, file

2. Database, file, record, field

3. File, record, field, database

4. Field, record, file, database

12. Four basic functions in manufacturing are:

1. Product design, marketing, accounting, and inspection

2. Business, engineering, production, human resources.

3. MRP–II, FMS, JIT, CAD

4. All the above.

5. None of the above.

13. The acronym for a global computer-integrated manufacturing system is:

1. CIM

2. CIMS

3. CAPACS

4. FMS

14. Historically, in a nontraditional manufacturing environment, the host computer was in an accounting world.

1. True

2. False

15. _____ and _____ systems form the hub of a CIM system.

1. Computers, machines

2. Procedures, computers

3. Communication, databases

4. All the above.

5. None of the above.

16. _____ should be actively involved in all CIM development processes.

1. Management

2. Employees

3. Support groups

4. Customers

5. All the above.

17. Two basic models used in database management are _____ and _____.

1. Partitioned, replicated

2. Relational, hierarchical

3. Vertical topology, horizontal topology

4. Network, plex

18. Protocol is a set of established rules that allows smooth and orderly transfer of information.

True   False

19. _____ planning involves forecasting the demand for the firm's products and services and translating this forecast into its equivalent demand for various factors of production.

1. Product
2. Process
3. Production
4. Work
5. All the above.

20. A CIM system must be carefully planned in advance by a multifunctional team approach.

1. True
2. False

21. The three basic operations of an information system are

1. Processing, input, output
2. Input, processing, output
3. Output, processing, input
4. Input, output, processing

22. Which is the best example of real-time processing?

1. Printing payroll checks
2. Summarizing daily transactions
3. Accumulating weekly inventory
4. Requesting an airline reservation

ANSWERS TO SELF-STUDY TEST

1.	True	**6.**	4	**11.**	2	**16.**	5	**21.**	2
2.	False	**7.**	5	**12.**	1	**17.**	2	**22.**	4
3.	True	**8.**	3	**13.**	1	**18.**	True		
4.	5	**9.**	2	**14.**	2	**19.**	1		
5.	4	**10.**	2	**15.**	3	**20.**	1		

References

[1] Gondert, Stephen J., "Understanding The Impact of Computer-Integrated Manufacturing." *Manufacturing Engineering,* September 1984, p. 68.

[2] Koves, Gabor, "Computer Integrated Manufacturing: An Overview." *Automotive Engineering,* Vol. 94, No. 6, June 1986.

[3] Harmonosky, Catherine E., "Integrating Automated Systems Control and Design." *CIM Review,* Spring 1986.

[4] Taylor, Alfred P., "The Factory With A Future." *Technical Paper,* Autofact 4 Conference, SME, Dearborn, MI, September 30, 1982.

[5] Brimson, James A., and Downey, P. J., "Future Technology: A Key To Manufacturing Integration." *CIM Review,* Spring 1986.

[6] Slattery, Thomas J., "Is CIM a Certainty?" *Machine and Tool Blue Book,* December, 1985.

[7] Weber, Thomas J., "The CAD/CAM Data Base . . . The Foundation of CIM," Commline, January-February, 1985.

[8] Froyd, Stanley G., "Relational Database Cornerstone of FMS." *Commline,* January-February, 1985.

[9] Sobczak, Thomas V., "A Glossary of Terms For Computer Integrated Manufacturing." *Computer and Automated System Association,* SME, 1984.

[10] _____, "Industrial Automation Productivity Issues: Plant Wide Communication." Allen-Bradley Systems Division, Highland Heights, OH.

[11] Baer, Tony, "Justifying CIM: The Numbers Really Are There." *Managing Automation,* March 1988, p. 30.

[12] Martin, J. B., "CIM: What The Future Holds." *Manufacturing Engineering,* January 1988, p. 36.

[13] Gunn, Thomas, "The CIM Connection." *Datamation,* February 1986, p. 50.

[14] Stauffer, Robert N., "General Electric's CIM System Automates Entire Business Cycle." *CIM Technology,* Winter, 1984, p. 20.

[15] Hegland, Donald E., "The Challenge of Computer Integrated Manufacturing." *Production Engineering,* October 1984, p. 52.

8

Planning for Computer-Integrated Manufacturing

Overview

As a global concept of manufacturing, computer-integrated manufacturing (CIM) is a complex operation, encompassing many key resources and the dedicated efforts of many individuals at all levels of the enterprise. The CIM concept means streamlining the product processes and support activities by bringing together and sharing manufacturing information from business, engineering, production, and human resources systems.

The implementation of a global CIM system translates into top management involvement, support, and efforts. As a result, CIM efforts require and demand effective planning and control from an initial CIM investigation, all phases in between, up to and including post-implementation review as shown in a conceptual CIM System Development Life Cycle (SDLC) (Figure 8.1). The SDLC shown in Figure 8.1 is conceptual only and is not a standard for all enterprises. The eight phases shown in the figure are critical to the success of an implemented CIM subsystem (CIMS). A CIMS is an automated system (or subsystem) with the potential of computer integration into CIM. Each of the phases in the SDLC must be carefully planned. Phases of the SDLC are discussed in Chapter 8 and Chapter 10. The approach of Deloitte Haskins & Selles to a planning cycle for CIM and a CIMS implementation is shown in Figure 8.2 [11]. This planning cycle is characterized as an iterative process at the two top levels of a traditional three-level (strategic, tactical,

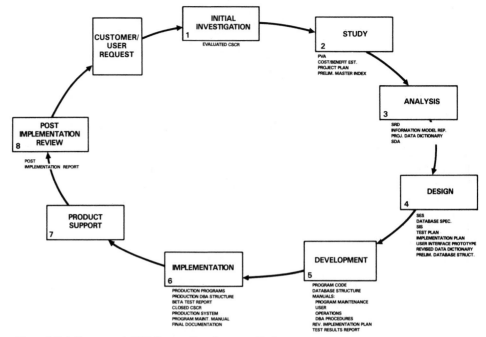

Figure 8.1 A Conceptual CIM System Development Cycle

and operational) triangle. The iterative process at the two levels consists of five identifiable steps or phases:

Strategic Level

1. MISSION: Definition. Highest-level vision of what the corporation/business, unit/division/enterprise will be doing.
2. STRATEGY: Approach. A plan, method, or series of actions employing all available resources of a company to achieve the stated mission with a 5- to 10-year planning outlook.
3. BUSINESS PLAN: Differential. Translates the company strategy into specific objectives and measurable goals within a 3- to 5-year time frame. Combines both the future and the operating plan for the next year.

Tactical Level

4. FUNCTIONAL STRATEGY: Action. Each major functional area (business, engineering, human resources, and production) develops and documents its plans, actions, and high-level requirements to support the goals and objectives of the company business plan.
5. CIM or TECHNOLOGY STRATEGY: Integration. A CIM strategy unifies the philosophies, information, and technology requirements from the individual functional strategies, and creates an integrated technology plan for the business.

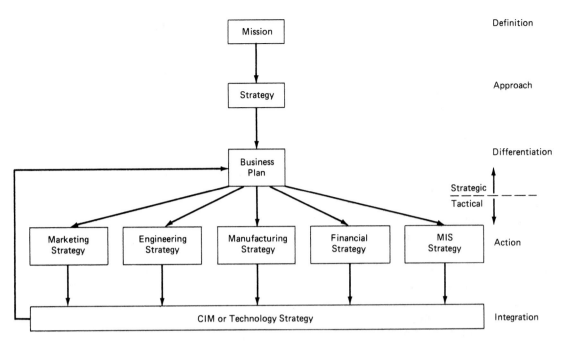

Figure 8.2 A Planning Cycle

A CIM planning cycle may vary from company to company. However, the CIM planning cycle shown in Figure 8.2 illustrates that a successful global CIM system involves the interrelationship of many functions striving to meet the requirements of the enterprise's business plan. Strategic planning for CIM requires teamwork throughout the company or enterprise. The team approach to CIM planning is discussed on page 279.

In summary, strategic planning for CIM involves the following:

- Top-down factory analysis
- Modernization roadmap
- Simplification and optimization Analysis
- Information system integration Planning
- Organization transition plan
- Standards and guidelines

The CIM Philosophy in Retrospective

We have referred to CIM as an information system for the complete manufacturing enterprise in which every aspect of the manufacturing process has been computerized and tied into a single, controlling, coordinating system. All applications of CAD and CAM, including management planning and control, are in the system. Data communications among different responsibilities in the management process

—from management decisionmaking to that of shop floor managers—is accomplished through shared access to the system's database files [1, 187]. A CIM system is not a technology but a way of organizing manufacturing technologies for a smoother flow of information, greater efficiency, and faster product development. Therefore, CIM strategies vary widely [2, 30]. Keep in mind that no matter what anybody tries to tell you, there is *no* single CIM system or solution.

The Impact of CIM

Large expenditures for computers and software, equipment, control and support systems are just for starters—there will also be big outlays for detailed evaluation and planning, for personnel training and retraining, and for development of support groups. A global CIM implementation process does not come cheaply. As a result, the CIM system and its implementation concept amount to a corporate revolution [4, 2].

Implementation takes careful planning and a step-by-step strategy. A company should not sit back and wait for the technology to stabilize. If so, the organization may not catch up with the competitor who is thrashing out the problems today. Every company that wants to be a significant worldwide competitor in the future must be willing to make investments today in money, time, and people to ensure the successful implementation of a CIM system [5, 69].

A first step during Phase 1 (Figure 8.1) in bringing CIM to a manufacturing environment is to answer two questions: (1) What is to be accomplished with CIM? and (2) Where should the company begin implementing CIM? These questions should be answered early in the planning cycle. Question 1 should be addressed at the first level of the interactive process discussed earlier in this chapter. Question 2 should be addressed by using the checklist that follows [7]. Action teams (discussed in Chapter 8) can be established to perform these tasks.

A Checklist for CIM

First, within your manufacturing business, consider the functional areas where implementing CIM would have the highest priority, second highest priority, and so on. Typical of such functional areas are the following:

- Design Engineering
    Drafting
    Documentation
    Project Management
    Analysis
- Manufacturing Engineering
    Engineering Change Orders
    Process Planning
    NC Programming
    Tool Design

- Production Planning
  - Plant Layout
  - BOM Scheduling
  - Machine Tooling
  - Manufacturing Cost
- Materials Management
  - Inventory Control
  - BOM Control
  - Parts Control
  - Tool Management
- Production
  - Work in Progress
  - Scheduling
  - Finished Goods Inventory
  - Configuration Control
- Purchasing
  - Vendor Performance Analysis
  - Receiving/Inspection
  - Contract Administration
  - Purchase Order Control
- Finance/Accounting
  - Accounts Payable
  - Accounts Receivable
  - General Ledger
  - Cost Accounting
- Data Processing/MIS
  - Database Integration
  - Communications
  - Development Tools
  - Hardware Compatibility

Within these functional areas, consider specific tasks that would benefit most from CIM.

Basic Strategies for CIM Solutions

Strategies for implementing all phases of CIM require more than creating an atmosphere that encourages systems integration. The strategies also offer challenges at all levels of the enterprise. Typical challenges for implementing CIM are given in Table 8.1, which highlights the areas of concern at the three management levels of the enterprise. These areas are: (1) motivation techniques, (2) know-how skills, and (3) necessary tools and organization. These concerns serve as a focal point for the planning processes.

TABLE 8.1 Challenges for Implementing Computer-Integrated Manufacturing Systems

SIZE OF COMPANY: ☐ Large ☐ Medium ☐ Small

Level	Motivation to Implement CIM	Know-how to Implement CIM	Tools/Organization to Implement CIM
Executive	Stay in Business Profit Sharing Company Image (Retirement Soon) (Short-Term Commitment) Personal Recognition National Concern (Risk) Financial Reward Expand Business	CIM Knowledge and Input System Architecture Computer Literacy Organizational Skills Team Building Find Models of Success Learn to Simplify Complex Systems Management of Interdependent Resources Cost Justification	System Design and Similation CIM Project Team Diagnostics System Performance Measuring System Outside Sources of Information

Mid-Manager and Technician	Personal Growth
	Increase Efficiency
	Improve Control
	Enhance Quality
	(Lack of Corporate Support)
	(Lack of Information Sharing)
	(Reward for Competition instead of
	  Cooperation)
	(Not Invented Here)
		Computer Literacy
		Systems Engineering
		Software Engineering
		Group Technology
		Modeling and Simulation
		Database/Communication
		  Standards
		Knowledge Engineering
		Algorithm Capture
		Ability to Develop Transition Plans
		Team Building Skills
		Feed-Back From Shop Floor
			Graphics Design Stations
			Active Data Dictionary
			Standards
			  Graphical Report Generators
			Documentation Systems
			Database Administrators
			Knowledge-Based Information
			  Standards Committee
			Interactive Training System
			System Design Tools
			Outside Sources of Information

Operator Craftsman	Profit Sharing
	Remain Employed
	Organized Operation
	Challenge/Satisfaction
	(Learn New Skills)
	(Eventual Job Elimination)
	(Greater Responsibility in Failure)
	Personal Image
	Authority and Responsibility
	Quality of Work Life
		Computer Literacy
		Increased Analytic and Operational
		  Skills
		Broadened Scope of Responsibility
		(Provide Time for Continued
		  Training)
			Interactive Training

Strategies for implementing CIM must provide an environment in which people can work together in teams, share ideas, and make contributions under a strong CIM team leadership. Many important activities must take place during the implementation of the functional areas on the checklist. These activities must be scheduled, coordinated, and managed in a timely way and must support the phases shown in Figure 8.1.

Organizing for CIM

An important strategy early in the CIM solutions process is the establishment of an organizational structure that reflects the three levels of management of the CIM entity. The structure should be headed by a chief executive officer (CEO) and vice CEOs who form the CIM executive management team (EMT) (Chapter 8). The CIM entity can be reflected in many forms. A typical example used in this discussion is a CIM center. An organization chart of a suggested CIM Center is shown in Figure 8.3. This center should be used as a centralized CIM structure, a team management hub for all CIM operations and activities. By having such centralized talent and being a component independent of the other functions, this CIM organization can perform the integrating role that is the heart of the global CIM concept. This aspect of the organization chart illustrates the CIM Center EMT and shows a structured functional relationship between the CIM program manager (CEO) and the CIM project managers (vice CEOs) of each major function discussed in Chapter 1. This is the highest level of team management; the CEO serves as the chairperson of the team. A special structured layer technique is used so that each major function may distribute responsibilities to managers of lower-level subfunctions and/or of CIM management teams. Such an arrangement is shown under business and human resources in Figure 8.3.

The CIM Center EMT in Figure 8.3 provides leadership in developing a master CIM plan and defining a basic implementation strategy. It should be well

Figure 8.3 Organizing for CIM

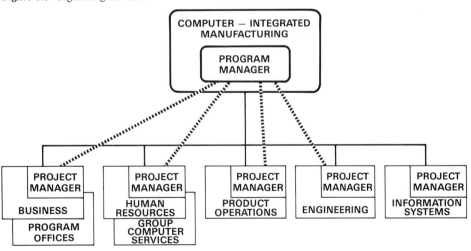

supported by CIM multifunctional teams (page 270). The team also provides an organizational transition plan which addresses the following:

1. Clearly stated common vision and goals.
2. Defined new roles (formal CIM organization and project or action teams).
3. Approaches to minimize conflicting incentives.
4. Plans to achieve and maintain "critical mass" (on-going education and training, cross-functional participation).

Zygmont has pointed out that there are three distinct strategies for implementing CIM [2, 28], as follows:

- Inside to Outside: This approach creates a computer network for data exchange between a manufacturer and its suppliers and/or customers.
- Beginning to End: This approach creates a product's development cycle, creating a data continuum from earliest design and planning through engineering and production, and even tying in support departments like marketing and technical publications.
- Top to Bottom: This strategy disseminates information downward for better control of manufacturing operations and to feed information upward from the shop floor for use in business management and planning.

The strategy used will depend on the mission and personality of the enterprise. In refining the CIM Center EMT's basic implementation strategy, considerations should be based on such factors as the following:

- Business goals and objectives
- Strategic business plans
- Management philosophies and approaches
- Size and organization structure of the enterprise
- Current and future facilities and plant layout
- Conceptual CIM architecture
- Management commitment and support
- Manufacturing's requirements for resources
- Human resources—education and training

A 10-point CIM checklist to assist the EMT with its planning processes is provided in Table 8.2 The items on the checklist are not all-inclusive but serve only as a guide. Items on a checklist will vary from enterprise to enterprise.

Because implementing CIM involves developing and integrating a host of subsystems (CIMs), one component of the team is charged with maintaining the "big picture." This component is the CIM Center EMT. The CIM Center EMT also has the responsibility of seeing that the subsystems (CIMs) of the major functions are integrated according to a well-developed plan, the end results of which are to

TABLE 8.2 Ten-Point CIM Checklist for Planning

1. **Strategic Plan**
   - Where are you?
   - Where do you want to go?
   - What are you willing to do to get there?
2. **Business Architecture**
   - What is current architecture?
   - What problems does it have?
   - What can you change?
3. **Cost Centers**
   - What are your current costs?
   - What is a competitive cost?
4. **Data/Knowledge Base**
   - Do you understand the CIM database requirements?
   - Do you have a computerized database?
   - Is it accurate/credible/complete?
   - Will you create or correct it?
5. **Communication System**
   - Do you understand what CIM requires?
   - Do you have a communication system to support CIM?
6. **Interface to Outside World**
   - Have you defined interfaces to the outside world such as customers, vendors, public relations, and legal?
   - Have you made plans to respond to these?
7. **Resources (human, materials, financial)**
   - Have you analyzed current resources in terms of capability and capacity?
   - Have you analyzed resources required for CIM?
8. **Processes, Equipment, Tooling**
   - Have you inventoried your capabilities?
   - What modifications do you need to support CIM?
9. **Expansion Plans**
   - Are your plans for introduction of CIM evolutionary or revolutionary?
   - Are your architectural models sufficient for your approach?
10. **Regulations and Standards**
    - Have you identified standards for CIM?
    - Have you analyzed the risks associated with use or nonuse of standards?

combine and integrate CIMs into a manageable CIM system. The new integrated system, consisting of old, new, and latest technologies, should provide significant improvements over the old existing systems. These improvements should be in terms of efficiency, flexibility, and reliability for the manufacturing enterprise. Additional information on this topic is discussed on page 274.

Management Commitment

The success of the CIM solutions planning process will depend on the support of management. Management at all levels of the enterprise should be involved in the planning process, as well as all CIM solutions efforts. The EMT organizational structure visibly shows commitment by top management. Improvement comes through a willingness on the part of top management to throw itself, along with resources throughout the organization, into the process. It also comes through a willingness of top management to tolerate change [3, 15]. Other levels of management teams are discussed on page 278.

One of the basic responsibilities of the executive management teams is to provide management with an understanding of the CIM concepts and the impacts these concepts will have on the product processes. Management's understanding of CIM may come from through direct involvement and through various CIM awareness and educational activities. Management should play a vital role in the educational process. None is as important as having the decisionmaking level of management directing some of the awareness and educational activities related to CIM projects and CIMs. This is another way by which management may show its commitment to efforts to find CIM solutions.

Educational Processes

In keeping with the philosophy of the global CIM concept, computer integration of manufacturing operations over a complete manufacturing cycle requires human resources to have special skills and requirements. Human resource development should be involved early in the CIM process on page 281 and Chapter 9.

In-service CIM awareness and educational activities are essential. A key ingredient to CIM solutions is based on the company's educational policy and how to make it work for the company. Serious problems are likely to arise from time to time if there is a lack of proper education and training. Human resources must address the following educational issues:

1. Specific types of technological skills needed in a CIM environment.
2. How to upgrade workers' technical skills to deal with new equipment and systems.
3. Where and how to find trained or trainable personnel in the enterprise.
4. A CIM awareness program for all levels of the enterprise.
5. How each employee can be a key player on the CIM team.
6. Implementation and evaluation of the company's CIM educational program.

Sponsorships

A CIM system is built on the foundation of many modular CIMs tailored specifically to the needs of various functions in the manufacturing enterprise. The modular CIMs are owned by these functions and are their responsibility. As a result, like CAD/CAMs, they require active sponsorship. Ness and Jacobson [6, 8–9] point out that successful sponsorship cannot be restricted to isolated, selected individuals but must pervade a complete chain of command. In large organizations with many chains and multiple projects, effective sponsorship emanates from a project's line management rather than from central staff management. A credible and influential project management chain whose natural risk aversion is outweighed by their conviction that predicted benefits can be realized must sponsor integrated CAD/CAM. For successful implementations, practical pressure is required to overcome impediments to change caused by the inertia that exists in the familiar ways of doing things.

Active sponsorship of a CIMs, coupled with some risk taking and visionary management, are required to set up the model project, which, when proven successful, will be expanded throughout the company.

Talent Sharing

Study teams should be created early to work on CIM solutions. These study teams, or action teams, should consist of representatives from all interrelated departments and/or functions as well as information systems, reliability and quality controls, and data management and communications (page 278). Within each team, a technical core group should be formed that will be responsible for technical evaluation of the techniques and vendors of CIMs. Team designs are discussed on page 275.

The management team approach is a unique way of sharing talents and information and of exchanging ideas between members on each team and between interrelated functional and departmental teams. By this means, timely multifunctional decisions can be made. As a result, every person in the enterprise plays a more significant role in the CIM implementation process. Each person becomes a decision point, a node, within a larger network of people who can work in parallel in an interactive manner.

Off-Site Meetings

A CIM system is an interdepartmental and multifunctional manufacturing business designed to produce competitively priced quality products in order to make a profit. Such an operation makes it difficult for a single person in one department to make good and sound decisions for people in other departments. It is also less likely that people in other departments will accept decisions, good or bad, from people outside their own departments. The personnel of one department are likely to be reluctant to make decisions that will affect the personnel of another department. It is therefore necessary to establish an environment in which key representatives from interrelated departments and functions can come together in a relaxed environment to discuss common issues and operations and make multifunctional decisions. One successful technique of making interfunctional decisions is through off site meetings.

Off-site meetings are usually very successful if well planned and if the plans are well executed. The meetings should not be conducted as if they are bringing the members back into their work environment. Instead, they should be conducted in such a way as to generate and evaluate potential CIM solutions. As a result, off-site meeting strategies will vary from meeting to meeting. Wittny [9, 103] points out that team meetings should be conducted with discipline but not with undue formality. The members must think through complex issues and not just vote on the apparent popularity of ideas. The meetings should be conducted with enough regularity to ensure that steady progress is made.

Team Management and Leadership Roles in CIM

Emphasis in earlier sections of this chapter was directed toward *strategies* and *organizing* for CIM solutions. This section addresses team leadership and responsibilities. A CIM system requires extremely close cooperation between individuals, groups, departments, functions, and management at every level of the enterprise.

This new way of doing manufacturing business may cause the enterprise to make radical changes in how it operates. To avoid poorly designed and integrated CIMs, CIM teams should be put together at every level of the enterprise with membership from interrelated departments and functions. Interfunctional CIM team building and operation require management commitment at each level. Management should make resources available so that the teams can be in an environment conducive to good work.

Introduction to Team Management

The implementation of CIM and/or CIMs should not use hit-and-run approaches to CIM issues, efforts, and solutions. Being a cross-functional operation, CIM does not lend itself well to many of the traditional management techniques. As a result, CIM should be and must be an enterprise strategy that employs new and refined traditional management techniques to successfully face its future manufacturing challenges. One such technique is nodal management in which each departmental function and subfunction become a *node,* or *decision* point. Computer and Automated Systems Association of Manufacturing Engineers (CASA/SME) refers to such a technique as the "Fifth Generation Management for Fifth Generation Technology" (FGM) [10].

Fifth generation management (FGM) is defined as follows: "The sharing of visions through communications and trust, based on honesty and integrity, accomplished by the integration of labor, the awareness of interdependencies, and management through the use of program management tools which achieves further enlightenment of employees in order to manage competitiveness" [10, 31].

This kind of management calls for a new leadership. It nurtures an integration in which the knowledge centers are expected to work together in an interactive fashion. It expects more of all employees to be involved in the decision-making process in dealing with a business variety so windows of business and technical opportunity can be more effectively met. It taps the power, wisdom, and insight of its managers, professionals, and employees who have, as a team in a nodal network, learned to use computer-based resources to enhance their decisionmaking capabilities. Human interaction in FGM is recognized as an important issue. Its shared vision, values, and culture coordinates FGM rather than its being held together by a command and control structure.

Such key issues as those that follow may be addressed and managed through FGM:

- Trust of superiors, peers, subordinates inside and outside of one's organization.
- Open communications between people and organizations.

- Removal of organization barriers.
- Teams consisting of knowledged-based individuals.
- People and work managed by using program management techniques.
- People given more responsibility and their experience and knowledge used.

When an opportunity is presented for developing and implementing a CIMs into the enterprise, it is of considerable value to form multifunction CIM teams. These teams become CIM "Management" and "Action" teams requiring guidance and management support in effecting changes in the method of operating the business and gaining the full benefit of CIM technology.

Multifunction Teams in CIM

Multifunction teams in CIM are of two general types: management and action. Multifunctional management teams focus on the management of the integration of most all aspects of the enterprise at all levels. The CIM support and commitment are discussed in Chapter 8. Multifunctional action teams are discussed below and in Chapter 3.

Action teams, like ad hoc groups, are mobilized to tackle specific problems. However, CIM action teams, unlike traditional ad hoc groups, are multifunctional teams and use FGM techniques to tackle the problems with representation from each of the involved functions. The reasons for multifunction teams are as follows [9, 92]:

1. Used as a decisionmaking body
2. Represents thoughts of all functions
3. Decisions are based on collective knowledge of many employees because nobody knows everything
4. Generates a high degree of cooperation between functions at all levels

The more complex a systems project is, the greater the responsibilities are. Typical CIM multifunction team responsibilities follow:

1. Resolution of procedural rules
2. Evaluation of mutual impact
3. Generation of solutions
4. Project justification
5. Progress monitoring
6. Installation planning
7. Communication and buy-in
8. Significant decisions
9. Resolution of conflict

Team Composition

The composition of an action team varies according to its function [10, 32]. Generally, the team includes representation from the functions that are likely to be affected by the proposed CIM system. The membership should include persons with knowledge of manufacturing operations, production control, facilities engineering, process engineering, factory planning, and information systems. A suggested guide for action teams membership makeup is the following [9, 33]:

1. All involved departments
2. Broadly experienced people
3. Data processing representative
4. Middle-level managers who know the system, make decisions, are able to compromise, and have influence within department

Other support organizations to the affected departments may be tapped as needed for additional contributions.

Qualifications of Team Leaders

Many issues must be resolved and many tasks must be performed during the CIM implementation processes. As a result, many teams must be used and guided under an effective team leader. A team leader may be defined as a person who oversees a task or project and guides the members on the team to a successful completion of the assignment. The team leader has special leadership skills. The leader should be able to do the following:

- Prepare, organize, and conduct effective meetings.
- Be a participative leader; be innovative, creative, and able to motivate people.
- Be respected by team members, knowledgeable of the territory with good comprehension.
- Have no ax to grind; be trustworthy; be able to compromise and make decisions.
- Be respected by peers in the organization structure and have a good functional relationship with them.
- Pay public recognition to the contribution of the team members.

In summary, team leaders should be selected on the basis of the following:

1. Leadership Style
   Task-Oriented
   Relationship-Oriented
2. Situational Control
   Leader/Member Relations
   Task Structure
   Position Power

Support and Commitment to CIM

Management and Employees Participation

Issues across functional lines that involve CIM must be resolved, tasks at all levels of the enterprise must be performed, and many resources throughout the enterprise must be made available during the phases of the CIM implementation. As a result, CIM commitments, support, and motivation by management, management teams, action teams, and human resources must be generated throughout the enterprise. These requirements must be at all levels of the enterprise. Efforts should be made to motivate employees early during the CIM process and seek their commitment to the program. From a team point of view, action teams should be established when needed under effective leadership at three levels of the enterprise (Figure 8.3). Figure 8.3 has been revised to Figure 8.4 for further discussion. Figure 8.4 shows three levels of management. As a result, for the CIM planning processes, there should be three levels of team management: executive management team (EMT), functional management team (FMT), and operational management team (OMT). The goals and actions of these management teams should be such that they provide the stimulus to generate interest and motivation, provide resources, and show management commitment to the implementation of a CIM program. These management teams bring together work groups (action teams) to support their responsibilities.

There are four basic ingredients for a successful action team building process: (1) define the team's mission, goals, and objectives; (2) select team leadership; (3) kindle interest for membership in the action team, and select team; and (4) attend

Figure 8.4 Levels of Management for CIM Solutions

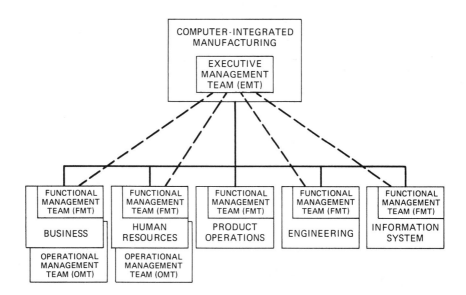

the initial meeting of the action team to assure it of management's commitment and support.

Roles of the CIM Management Teams

A CIM management team, through the use of various resources, should prepare a statement of purpose, its mission, goals, and objectives. The role of each action team should be clearly defined with an effective communication and feedback system.

Executive Management Team

The EMT is a CIM strategic management team. The EMT should be under the leadership of a CEO (page 269) who has the responsibility of guiding the enterprise toward the common goal of the CIM system. The EMT should be highly visible in CIM actions of the subordinate management teams. It should also be active in setting the proper environment for a CIM program, providing encouragement, and assuring management's commitment and support for the CIM program.

Functional Management Team

The FMT is a CIM tactical management team. The FMT should be under the leadership of a major functional manager. The FMT has the responsibility of guiding each major function to the common goal of CIM. It openly supports the CIM program. This level of team manager assigns the most capable persons to CIM projects, selects leaders of action teams, and is active in the process of building the team.

The FMT provides a communication and feedback channel up to the EMT, down to the OTM, and to all CIM action teams. The FMT is highly visible in interfunctional issues, such as communication, database, operational, and multifunctional actions. It has a good perspective on the automated systems of the various departments and their relationship with neighboring departments. As a result, FMT assures the action teams, departments, and functions that management is solidly committed to the CIM program.

Operational Management Team

The OMT is the third layer in the team management structure. One of the primary responsibilities of the OMT is to motivate people at the operational level and get everyone involved. Other involvements are to seek suggestions for improving specific tasks or systems, to encourage employees to become members of action teams, and to allow for them to become team leaders, to provide training on new CIMs before and after implementation, and to participate in the formation of action teams.

The OMT recommends action team membership, project leaders, leaders of CIM action teams, CIMs, operations and systems improvement ideas, and CIM operational planning information to the FMT.

There are no generic CIM action teams performing a variety of CIM-related tasks. Action teams vary in purpose, mission, membership, size of the enterprise, and enterprise organizational structure. The business objectives and overall strategic plan of the enterprise along with a CIM strategic plan are the driving forces for the types of CIM action teams, when they will be initiated, and the life cycle of each. Pearson of AT&T Technologies points out that when the climate is right, an effort should be put forth to generally define what goals and benefits are expected by introducing CIM. When some idea of what is desired has been formulated, a multi-disciplined team should then be formed to further define the goals and generalize on how the goals can be accomplished [8, 31].

Pearson further states that in forming the team, members with a variety of training and experience should be included. Using a single discipline can seriously limit the effectiveness of the team in developing plans that will address all aspects of operating the business. The actual number of participants will vary with the complexity and size of the business and its objectives. It would not be unreasonable for a team to consist of two to five persons in a small company to a dozen or more in a substantial business.

Forming Action Teams

Action teams should be initiated early in the CIM system development life cycle (SDLC), as shown in Figure 8.1. This action team should be designed to assist the CEO and EMT with their responsibilities. Other action teams should be created as needed. Action teams are used throughout CIM system SDLC for such CIMs provided by the checklist. The EMT is the centralized management hub of all the teams.

System Development Life Cycle

One solution to team management is a centralized CIM management structure, as illustrated in Figure 8.3. This structure is organized for team management along the lines of the eight phases in the SDLC processes: (1) initial investigation, (2) study, (3) analysis, (4) design, (5) development, (6) implementation, (7) product support, and (8) post-implementation review. Each phase may have one or more action teams involved in its function. A brief description of each phase to be used as a guide in building the action team is discussed. Activities are not limited to those discussed for each phase. An SDLC is discussed in detail in Chapter 10.

Initial investigation. An initial investigation begins when the EMT receives a CIM system change request (CSCR), which is the formal mechanism for requesting EMT effort in system development. It lists the needs and requirements for the system development and potential benefits. As do all other action teams, the team that initiates the CSCR reports to the EMT.

Study. A Project Viability/Feasibility Assessment (PV/FA) document is prepared during this phase. It defines project scope and assumptions taking into account management, imposed constraints, budget limitations, and technical considerations involving other systems under development or planned for the future.

Analysis. The analysis function determines and lists the precise relationships among the components of a complex matter or system, and breaks them down into simpler components or elements. During this phase, a system requirements definition (SRD) document and a system design alternatives (SDA) document are prepared. The SRD document includes a description of the present system, a definition of requirements to be met by the proposed system, a list of anticipated benefits, and the recommended scope and schedule for work to be done next. The SDA document provides a synthesis of solutions for a given set of requirements and problems. It assesses the viability of various proposed solutions to the requirements, defines system design objectives for the selected alternatives, and includes hardware, software, and communication architecture.

Design. The design function prepares the plans for building a product or system. System external specifications (SES) and system internal specifications (SIS) are prepared during this phase. The SES provides a complete statement to the technical designer of what the user needs and expects the system to be and provides the user with an understanding of the inputs, outputs, and functions of each automated system. It also provides the general design specification that the user signs off as a go-ahead before the detail design and coding take place. The SIS defines how to build the system described in the SES. It identifies required programs, defines modular structure of each program, and specifies report layouts, record and file structures, and the like.

Development. Development consists of designing the product, testing the product, and evaluating the product. The program code, database structure, test results, and manuals including program maintenance and database administration (DBA) are developed during this phase.

Implementation. The implementation phase involves installing a system into operation. It concludes the system project with the exception of appropriate follow-up, or post-installation, review. It includes production programs, production DBA structure, Beta test report, closed CSCR, production system, program maintenance manual, and final documentation. (See Chapter 10 for a full discussion.)

Product support. On-going system support is provided during the product support phase. The user organization has the primary responsibility for operating the system, educating and training the operators, maintenance, and upgrading the system as required. This responsibility includes charges, evaluation of new ideas for enhancement, functional additions to the system, and feedback information for system improvement to the EMT.

Post-implementation review. A review of a CIMs and what has been accomplished after approximately six months of operation should be part of the post-implementation review process. A review is made of the development cost, operation costs, performance of the system, value of the system, improvements needed in system development techniques, and evaluation of the training for system operators.

Summary of Team Management

A successful global CIM system depends on such factors as management commitment; employees' motivation, interest, and participation; education and training at all levels of the enterprise; and a well-developed strategic plan. Strategies for carrying out the plan must be kept on track. One course of action is team management with interest at the enterprise level rather than the functional level.

The management teams should be multifunctional with representation from the functions most likely to be affected by the implementation of CIMs in the global CIM environment. Each team becomes a node (decision) in CIM. These nodes become reference points (knowledge centers) capable of teaming with other nodes in the support of the enterprise's CIM strategy.

This approach to planning and controlling CIM was placed in three levels to reflect the management levels of the enterprise. This technique provides for greater functional participation, controls, and integration.

Understanding CIM

Education and training programs for managers, employees, and support groups should be a vital part of the CIM implementation process. The types of programs and methods of presentation vary from group to group. One approach to this problem is to form a CIM action team on education early in the CIM planning process. The primary mission of this team should be to address educational issues. As previously stated, the use of action teams is one method of getting more people involved in the CIM processes. Such an approach to CIM educational issues can motivate, spark, and maintain interest in CIM among the employees. Brauninger listed the following techniques used to spark interest in CIM at the Cone Drive Company [8, 23]:

1. Developed expectations that things were going to get better—that new systems were going to be better than the old ways of doing things. This got all employees a little excited. They knew that the way things were done was not the best way but had never been given the tools to do a better job.

2. Trained employees in the new system before the actual installation. This developed confidence that they could do their jobs using more advanced systems and equipment. People were also promised that they would be trained in a new (and probably better) job if their current job was eliminated. This gave them security and a reason to look for time savings.

3. Remained open to suggestions for improvement whether it related to specific tasks or the system as a whole. Although the basic jobs needed to be done, the company was able to modify many tasks to be done the way the operator wanted.

4. Allowed people with ambition and desire to lead in the implementation of new ideas and projects.

5. Showed everyone the results as projects were successfully completed.

The established action team should be chartered to identify educational techniques, programs, and methods of delivery that would provide professional growth as well as introduce CIM concepts, impacts on the enterprise, and benefits to the enterprise and all employees. Obviously, if proper approaches are taken during early phases, interest in CIM in all operations at all levels of the enterprise will be generated. Amrine [12, 230] points out that a sound program of education and training will materially influence the success of any control program. This means education and training of all, from top management personnel down to the line worker. Of course, more detailed and advanced education and training are indicated for those personnel who will actually design, plan, develop, and operate the system. Again, this is not necessarily a one-time affair. Control systems maintain their optimum effectiveness only when a continuing program of education and training is conducted.

In-Service Education and Training

Functional areas in the enterprise that will most likely be effected severely as a result of CIM or CIMs implementation should be identified and addressed. The kinds of programs that will give support are discussed.

CIM Awareness

A CIM awareness program should be initiated early in the CIM implementation process. There are several reasons for doing so, among them the following:

1. People will have a tendency to turn away from things they know little about. The more they know about CIM and what it will do for them, the more likely is it that a base for their support will be generated.

2. Teach CIM concepts at all levels of the enterprise.

3. Overcome the normal opposition to the introduction of new CIMs.

4. Develop a sense of personal involvement in the CIM process by the employees.

5. Capture the types of educational programs that would maintain employees' interest in CIM.

6. Develop career counseling that will be provided throughout the CIM development processes.

A CIM awareness program is also a key to generating interest and enthusiasm in CIM and gaining the support of management and employees. The focus of this program should be to get people involved and thereby make them more receptive to the CIM development processes.

Retraining

An on-site retraining program is essential in CIM. It must be an on-going program to keep employees updated on CIMs. Lack of such a program can mean a failure for CIM. Vail points out that a company must staff the factory with workers who have the necessary qualifications [13, 252]. To do this, the company has two major options: (1) fire all present workers and hire all new ones, or (2) retrain the current workers it has. Obviously, the first option would be a setback, or even a disaster, for CIM. The political, legislative, workers attitudes, and labor relations groups uproars would be too great for the enterprise to implement CIM successfully if option one is taken.

The second option should be used, and there are many advantages in using it. Typical of them are that preference will be given to current employees rather than seeking qualified personnel from outside the company; good public image and labor relations will be established; the workers' support for CIM will be won; promotional opportunities for workers will be created; the upgrading of skills of current workers will be provided; and management will show support for job security, happiness, and improved productivity of its workers.

Training

On-site early training programs should be provided for new and current personnel on modern CIMs. Such programs accelerate these employees' learning and make them more productive earlier. They also free senior members in the enterprise to concentrate on more responsible tasks, stimulate employees' interest and involvement in CIM activities, provide exposure to management in CIM activities, and create for the employees promotion opportunities and an environment that is receptive to change.

Off-site training should be conducted in varying degrees to adequately prepare the key operating personnel before equipment is installed [14, 175]. Formal classes are also assets to the training program.

Professional Development

A professional development program provides educational experiences for professionals that enable them to keep abreast of changes in their fields. This program should also be designed to develop and strengthen the capabilities of professionals engaged in CIM and CIM-related activities. High-quality professional education may be provided through off-site seminars, professional meetings, courses at colleges and universities, exchange programs between industry and educational institutions, and the like.

On-site professional education should also be provided for the CIM professional. Many companies offer quality programs by using their own professional resources and also bringing in professionals from colleges and universities, consulting agencies, and so on. Management should provide means for the professionals to participate in on-site programs, and the employees should take advantage of these means to improve their abilities as CIM professionals. Those who keep up and who take every opportunity to improve themselves professionally prepare themselves for promotions, leadership positions, and salary increases.

Communications and News Reporting

Communication is a process whereby information may be transmitted and reported with understanding. Information must flow between levels of the enterprise and to and from various programs, functions, and operations. To link elements in the enterprise together, a good communications mechanism with feedback should be developed and used. Typical of CIM-related information is that related to new projects, successful operating projects, skills required for projects, training programs, concepts of CIM, CIM successes, employee awards for CIM accomplishment, and so on.

Techniques for passing this information throughout the enterprise may appear in many forms. One basic reporting form is the news report, which can be either *internal* or *external.*

Internal

Internal news reporting is a popular method of informing management, employees, and support groups of activities in the enterprise. It is an effective method of letting everyone in the enterprise know what is going on and how each function fits into the picture. Typical of techniques are bulletin boards, weekly or monthly status review meetings, newsletters or bulletins, and periodic business review meetings. Each technique should be carefully planned and properly implemented for effective reporting.

External

External news reporting overlaps internal reporting in some way and is used primarily to report the enterprise's successes and accomplishments to the general public. Techniques used to promote external reporting are press releases, technical publications, participation in professional organizations, mass communication, and speakers at professional conferences, educational institutions, and chambers of commerce.

Internal and external methods of reporting are effective methods of getting the word out about CIM, CIM activities, and the effects CIM has on improving manufacturing productivity. All kinds of communications are powerful tools for management to use to communicate with all employees. Communications play an extremely important role in planning, implementing CIMS, and policy-making phases of the management CIM process. Effective communications break down barriers between people, functions, and operations; they show management support for CIM; they generate interest and support for CIM.

A CIM Master Plan

A CIM master plan (CIM plan) should be developed around the premise that "CIM is not a technology. . . . It is a huge information system derived from the application technology for the integration of systems in a manufacturing enterprise." Several disciplines are involved in the development of this information system, including the following:

- A multifunctional team development concept.
- A new method of defining and selling the system.
- Innovative methods of designing, developing, implementing, and testing the new systems.
- Methods of maintaining project control over a new CIMs in a CIM environment.

Even though a CIM plan deals with technology, it is not a technical document. A CIM plan is a key element of a strategic and operational business plan. The CIM plan is a tool used to sell CIM and to educate management and employees about how the manufacturing business operates and how it could operate [8, 48].

A CIM plan points out the risks and impacts of CIM on all operations, functions, and management at the outset. It provides realistic expectations of what resources will be needed to computer-link product information of the entire enterprise, as well as where and when they will be needed.

A CIM plan should be prepared by a multifunctional action team under the leadership of a multifunctional management team. The plan should include, but not be limited to, the following areas:

- Scope of CIM efforts.
- Implementation principles and programs.
- The CIM subsystems funding, implementation procedures, and time schedules.
- Justification for CIM and each CIMs.
- A CIM proposal document.
- The "as is" business systems and the "to be" systems of the CIM plan.
- Common operating principles throughout the development cycle as well as ongoing operations.

Justification of CIM

Justification of CIM and CIMs is an integral part of the CIM planning process. Installing CIM or a CIMs such as MRP–II, FMS, CAD, CAPP, and GT in a CIM environment is one of the most lengthy, expensive, and complex tasks an enterprise can undertake. In some cases, an elementary CIMs may require more than two years to install and implement and can cost in excess of $3 million. The implementation and installation time for a more complex CIMs can be twice that of an elementary system and can cost up to 10 times as much.

CIM Subsystems (CIMs)

Justification of CIM projects is a major stumbling block in implementing a CIM strategic plan of the enterprise. Kaplin [15, 13] views a CIMs justification process as being caught in a strangling web of accounting and financial procedures, precedents, and regulations (many of them policies the firm itself instituted). Managers seem unable either to circumvent the traps or to adequately satisfy their requirements in order to proceed with the plan. Meredith [16, 242], of the University of Cincinnati, points out that this web comes about largely because automation is frequently driven by hardware, and hardware has historically been justified on the basis of bringing about either cost reduction or an increase in capacity—both requiring accounting verification. However, the benefits of these new automation technologies often have nothing to do with costs or capacity. Therefore, they stumble in the unnecessary pitfall of cost justification.

CIMs Justification Factors

Numerous factors justify a CIMs. They are generally one of two classes: noneconomic factors or economic factors. Both classes of factors should be included in a justification proposal—not economic factors alone. The benefits of CIM technology are real, quantifiable, and easy to identify. A number of new approaches are helping to justify CIM projects by using analytic as well as strategic techniques and procedures such as team management, action teams, motivation, and CIM strategic plans.

Noneconomic Factors

Many noneconomic benefits may be considered as cost benefits, and they play important roles in the justification of CIM projects. Many benefits can be identified, synthesized, analyzed, evaluated and presented to the CIMs justification process. The most common noneconomic factors are the following:

- Improved productivity.
- Reduced overhead expenses.
- Reduction of scrap, rework, and waste.
- Improved product quality.
- Competitively priced products.
- Improved system throughput.
- A reduction in the manufacturing cycle.
- The advancement of technology.

These cost benefits all add up to big savings and marketing advantages. The savings are in operating costs, working capital levels, and return on assets employed.

Economic Factors

Harrington stated as early as 1973 that justification must be a matter of conviction and not a matter of accounting [17]. Put another way, justification is a policy decision and not an investment decision. It is true that substantial investments are involved, and the financial resources of the corporation must be carefully considered, but these factors will govern the rate of investment, not the decision to invest. Economic factors are generally in one of the two categories: cost avoidance or cost savings.

An economic analysis of cost avoidance [18, 129] is made to determine the least costly of several alternatives. For example, a comparison may be made between a labor-intensive approach, special-purpose automation, and a robot CIMs. The choice to be made is not whether to retool the job but which approach to retooling will result in the lowest lifetime cost. The labor-intensive approach requires less capital outlay and has a relatively low obsolescence cost but results in higher operating expenses. The special-purpose approach requires relatively high capital expenditures and has moderate to low operating expenses with high obsolescence and changeover costs. The robot CIMs approach also involves a relatively high initial capital outlay and low to moderate operating expenses, but it has lower obsolescence and changeover costs than does special-purpose automation. An analysis of lifetime costs and expenses for each of these alternatives would be prepared and the least costly method chosen.

Cost savings, in contrast to cost avoidance where a capital expenditure is mandatory, apply in situations where there is the option to do nothing [18, 130]. In the cost savings case, the potential economic gain, as measured by return on investment (ROI) or payback period (PP), usually has to meet or exceed some established level to qualify for consideration.

Return on Investment

An example of an ROI analysis of a CIMs follows [19, 51]. The ROI formula is:

$$ROI = \frac{S \times 100}{T}$$

S = Total annual savings is determined by the formula:

$$(L + M - O) \times H - D$$

L = Direct and indirect labor		$20.00/hour
M = Material cost		1.00/hour
O = Maintenance and operation cost		2.00/hour
H = Hours of use during the year		2,000/hours
(one eight-hour shift)		
D = Annual depreciation		$10,000

$$S = (20 + 1 - 2) \times 2,000 - \$10,000$$
$$= \qquad 19 \times 2,000 - \$10,000$$
$$= \qquad 38,000 - \$10,000$$
$$= \qquad \$28,000$$

T = Total investment is determined by the formula:

$$T = P + A + I - C$$

P	= Purchase price	$60,000
A	= Accessories & extra equipment	30,000
I	= Engineering & installation costs	20,000
C	= Tax credit @ 10% of purchase price	6,000

$$T = 60,000 + 30,000 + 20,000 - \$6,000$$
$$= \qquad 110,000 - \$6,000$$
$$= \qquad \$104,000$$

$$ROI = \frac{S \times 100}{T}$$

$$= \frac{\$28,000 \times 100}{104,000}$$

$$= 26.92\%$$

Using the robot for two shifts per year, or 4,000 hours, changes the ROI favorably. For example:

$$S = (20 + 1 - 2) \times 4,000 - \$10,000$$
$$= \qquad 19 \times 4,000 - \$10,000$$
$$= \qquad 76,000 - \$10,000$$
$$= \qquad \$66,000$$

$$ROI = \frac{\$66,000 \times 100}{104,000}$$

$$= \frac{6,600,000}{104,000}$$

$$= 63.46\%$$

Payback Period

An analysis of the payback period of a CIMs follows [18, 140]. The payback period (PP) formula is:

$$P = \frac{I}{(L - E)}$$

P = Payback period in years
I = Total investment, CIMs & accessories $35,000
L = Total annual labor saving (including fringes)   16,000
E = Expense of CIMs upkeep (one-shift use)   2,000
                          (two-shift use)   3,000

One-Shift Use

$$P = \frac{I}{(L - E)} = \frac{35,000}{(16,000 - 2,000)} = 2.5 \text{ years}$$

Two-Shift Use

$$P = \frac{I}{(L - E)} = \frac{35,000}{(32,000 - 3,000)} = 1.2 \text{ years}$$

Justification of a CIMs or CIM project involves noneconomic factors that may be viewed as cost benefits. Economic factors are also used for justification and can be measured in several ways including ROI, payback period, and depreciated cash flow. Successful justification requires consideration of *all* the potential economic factors as well as the noneconomic cost benefits.

Guidelines for CIM Justification

- Look at the total picture.
    Identify alternatives and options.
    Evaluate consequences for not investing in CIM.
    Appraise your competitors' efforts.
    Define market opportunities opened up by CIM.
- Select a sufficiently long planning horizon.
    Five to ten years
- Conduct planning at a sufficiently high level.
    Management staff
    Interdisciplinary
- Consider all inputs to the financial justification.
    Estimate life cycle costs.
    Estimate working life of facilities.
    Estimate time patterns of all revenues and savings.
    Assign values to intangibles.
- Use discounted cash flow techniques. Recognize the limitations.
    Bias of discounting at high hurdle rates.
    Implied value of the status quo.
- Incorporate realistic risk assessments.
    Include actions that reduce risks.
    Estimate time patterns of risks.

The CIM Proposal Document

The CIM proposal document is an important part of the justification process. Its development must begin long before preparation of the actual document called for in a CIM master plan. The preproposal prepares management, employees, and support groups culturally and psychologically for CIM and CIMs technology.

Education for CIM (at all levels of the enterprise), internal communication systems, and professional activities are popular ways of motivating people to accept CIM, creating a CIM environment, and removing barriers to CIM. Removal of certain barriers opens doors for introducing CIM concepts and selling the benefits of CIM. A key element in proposing CIM is preselling it as thoroughly as possible before the work of developing the actual proposal is begun.

Proposal Document

A format for a CIM proposal and elements of the proposal will vary from enterprise to enterprise. Regardless of the diversity of the proposal elements, certain common elements should be included in any CIM proposal. Ewaldz and Hess list six common elements [8, 46].

1. Executive Summary. The proposal must start with a well-written executive summary. This is essential to win real commitment of top executives. No one can fully support something with only a vague idea of what it is all about. Executives must have a crisp, clear picture of the computer integration concept. The executive summary may be the last chance the proposal has to win over top executives. It may also be all they have time to read. Therefore, it had better be good.

2. Definition of the Computer Integration Package. The proposal should clearly define the overall structure of the computer integration package. It should identify the various modules—both hardware and software—that will comprise the system, and how the modules interface with one another. This description should act as a control and should be in enough detail so that any of the modules can be changed without requiring significant changes in the other modules.

3. Organization and Recommended Operating Policies and Procedure. The proposal should describe the organization and recommended operating policies and procedure under the computer-integrated system. This description must provide documentation for the changes needed and for the sources of the values that justify the investment. It is necessary to document the commitments made during the proposal development process, so that all persons involved can recall what changes need to be made to capture the benefits.

4. Implementation. The proposal should define in detail the implementation steps leading from the current environment to a fully computer-integrated firm. For a number of reasons, no organization should attempt to carry out the entire task at one time. The first reason is that the conversion is simply too drastic a cultural change for an organization.To try to make it in one "bite" will guarantee failure.

The second reason is that almost inevitably some elements of the technology are not going to work well at first, and it is a lot easier to troubleshoot problems one at a time than lots of them at once. Third, breaking the total investment into a series of smaller increments makes the investment more palatable, and in addition, allows the installer to develop skills and credibility implementation proceeds. Finally, computer technology is developing at an exponential rate; incremental implementation gives the firm a chance to take advantage of advances.

5. Design Guide for First Steps. The proposal should define the actions to be taken in the first steps. These actions should be in detail because they will act as the design guide for the first steps. This phase should define in detail the hardware needed, the software and operating system, the effort required to implement, and the changes required in the structure and organization of the firm. Specific suppliers can be identified, if appropriate. The proposal should also include detailed operating budgets for all functions involved, reflecting the improvements expected.

This level of detail is appropriate for the first few steps only, however. The designers and action teams of computer integration must have freedom to alter as necessary and practical. In the later steps, less and less detail should be included, but it should also be made clear that the risk of not achieving forecasted improvements will also become greater.

6. Financial Analysis. The proposal should conclude with the financial analysis of the proposed system, including a long-range—perhaps five years—projection of the costs and savings leading to a summary statement of justification and the payout period. The guidelines described in Chapter 8 should be reviewed carefully to help establish the case for this justification.

Summary of the Planning Process

Poor planning for CIM is one of the major contributors to unsuccessful implementation of the system. Systems do not function well unless they are carefully thought out and planned. The planning process must take into consideration the interdependencies of the various functions and operations within the manufacturing enterprise. At the same time, the functions and operations must work together through common procedures to accomplish the business strategic plan. To accomplish these two tasks, specific rules of operations and measures of performance should be developed for the individual operating functions to ensure cooperation within the CIM global system.

The thrusts of the planning process should be to:

1. Get everyone involved.
2. Set a CIM climate through structure and planning.
3. Form CIM management teams.
4. Form CIM action teams.

5. Sell the CIM Plan to management.

6. Promote a climate in which operations and functions can work together.

7. Establish good educational programs.

A CIM plan should define implementation principles, practices, and programs, as well as provide a reference for management to see how a CIMs fits into the big picture . . . a total CIM system. As a result, management at any time gauge the impact of a CIMs on other subsystems and the entire business enterprise if one CIMs is accelerated, delayed, or scrapped.

Exercises

Homework Assignments

1. Define team management. Explain how team management is used in the implementation of CIM.

2. Explain the difference between CIM and CIMs.

3. a. What is an action team?

   b. Explain the role of an action team.

   c. What should be the membership composition of an action team?

4. Explain the difference, if any, between team management and action teams.

5. a. What is a CIM plan?

   b. Explain the difference, if any, between a CIM plan (CIM master plan) and CIM proposal.

6. Is CIM a technology? Explain.

7. Assume that you have been elected as the team leader of an action team on "Organizing for CIM." Develop an organizational structure you would present to the team for consideration.

8. a. What is a multifunctional team?

   b. Explain the difference, if any, between an action team and multifunctional team.

9. List some techniques you would use in providing educational experiences for (a) manufacturing professionals, (b) employees needing retraining, and (c) CIM awareness programs.

10. List some advantages and disadvantages of off-site meetings.

11. List and define the major phases of a CIM system development life cycle (SDLC).

12. Define (a) CSCR, (b) PVA, (c) SRD, (d) SES and SIS, (e) SDA, and (f) DBA.

13. Explain the difference between economic factors and noneconomic factors as justification for CIMs.

14. Explain how you would involve a CEO on an executive management team (EMT).

15. What functional areas should be represented in a team effort for definition, design, and implementation of a multifunctional information system effort?

16. Prepare for training sessions for newly hired interactive graphics operating personnel.

17. You are a member of a planning group in an electronics manufacturing plant making integrated circuits. What type or types of CIMs would you propose for the plant?

18. Every team must have a leader if it is to be effective. List some qualifications for the leader of the team.

19. Explain the difference, if any, between Fifth Generation Management Functional Team Management, and Nodal Management.

20. List and discuss the needs for a multidisciplinary or multifunctional management team.

21. (a) List several levels (overlays) of team management. (b) Compare the responsibilities between each level listed.

22. List and discuss some guidelines for successful team meetings.

23. You have been given the responsibility for the development of a CAD/CAM database system that will impact a number of different departments within the production cycle. (a) List your functional team membership, (b) show an organizational structure for this functional team, and (c) show multiple levels of management for this structure.

24. List some basic operational guidelines for the team identified in Exercise 23.

25. Plan the agenda for the initial planning meeting in Exercise 23.

Self-Study Test

1. The primary method to spark interest in CIM is to get _____ involved.
   1. CEO
   2. Functional Management
   3. Management Teams
   4. Employees
   5. All of the above

2. In forming a management team, all members selected should have a variety of training and experience.

True   False

3. Typical systems team leader qualities are:

1. Hard worker, respected by team members, knows everything about CIM, good decision maker.

2. Good leadership skills, respected by team members, good at generating "axe grinding techniques," able to compromise.

3. Highly placed in the organization, good leadership skills, good systems comprehension, respected by team members.

4. #1 and #3 above

4. Important CIM implementation challenges area(s) of concern at the three management levels of the enterprise is (are):

1. Motivation techniques

2. Know-how skills

3. Necessary tools and organization

4. All of the above answers

5. None of the above answers

5. An important "early" strategy in the CIM solutions process is the establishment of an organization structure that reflects the three levels of management of the CIM entity.

True   False

6. A CIM master plan (CIM plan) should not be developed around the premise that CIM is a technology.

True   False

7. The justification of CIM projects is a major stumbling block in implementing a CIM strategic plan of the enterprise.

True   False

8. Common noneconomic justification factors are:

1. Improved system throughput, the advancement of technology, improved productivity.

2. A reduction in the manufacturing cycle, reduced overload expenses, improved product quality.

3. Competitively priced products, improved product quality.

4. #1 and #3 above

5. All of the above answers

9. A format for a CIM proposal and elements of the proposal will vary from enterprise to enterprise.

True   False

10. Poor planning for CIM is one of the major contributors to unsuccessful implementation of the system.
    True   False

11. CIM team meetings should be conducted with discipline, but not with undue formality.
    True   False

12. CIM is a narrow, highly specialized, all-encompassing concept.
    True   False

13. A management team is a decision-making team with representation from each of the involved departments that are most likely to be affected by the implementation of a CIM.
    True   False

14. The heads of departments are usually the most qualified persons to be team leaders.
    True   False

ANSWERS TO SELF-STUDY TEST

1. 5	**4.** 4	**7.** 1	**10.** 1	**13.** 1
2. 2	**5.** 1	**8.** 5	**11.** 1	**14.** 2
3. 3	**6.** 2	**9.** 1	**12.** 2	

References

[1] Allen, David C., "Techniques: Computer-Integrated Manufacturing." *Datamation,* March 1980.

[2] Zygmont, Jeffrey, "Manufacturers Move Toward Computer Integration." *High Technology,* February 1987.

[3] Rhodes, W. L., Jr., "Pulling It All Together." *Manufacturing Systems,* Summer, 1984.

[4] Slattery, Thomas J., "Is CIM a Certainty?" *Machine and Tool Blue Book,* Special Edition, 1985.

[5] Gondert, Stephen J., "Understanding The Impact of Computer-Integrated Manufacturing." *Manufacturing Engineering,* September 1984.

[6] Ness, P. H., and Jacobson, H., "Integrated CAD/CAM: A Case History and Management Perspective." *AUTOFACT 4 Conference Proceedings,* CASA/SME, Dearborn, MI, 1982.

[7] Prime Computers, "Computer Integrated Manufacturing: CIM by Design." *Prime Computer,* BR 1314, 1986.

[8] Chiantella, Nathan A., "Management Guide for CIM: Computer-Integrated Manufacturing Management Teams Report Steps for Successful Business Strategies." CASA/SME, Dearborn, MI, 1986.

[9] Wittny, Eugene J., "Managing Information Systems: An Integrated Approach." Society of Manufacturing Engineers, Dearborn, MI, 1987.

[10] Society of Manufacturing Engineers CASA/SME Technical Council, "Fifth Generation Management for Fifth Generation Technology." *SME Blue Book Series,* Dearborn, MI, 1988.

[11] Deloitte Haskins & Sells, "Computer Integrated Manufacturing Workshop." Hughes C & DP Institute, Woodland Hills, CA, January 1988.

[12] Amrine, H. T., Ritchey, J. A., and Moodie, C. L., *Manufacturing and Organization Management,* 5th ed. Prentice-Hall, Englewood Cliffs, NJ, 1987.

[13] Vail, Peter S., *Computer Integrated Manufacturing.* PWS-Kent, Boston, MA, 1988.

[14] Preston, E. J., Crawford, G. W., and Coticchia, M. E., *CAD/CAM Systems: Justification, Implementation, and Productivity.* Dekker, New York, 1985.

[15] Kaplin, R. S., "Yesterday's Accounting Undermines Production." *Harvard Business Review,* Vol. 62, July–August 1984.

[16] Drozda, Tom, *Flexible Manufacturing Systems,* 2nd ed. SME, Dearborn, MI, 1988.

[17] Harrington, J., Jr., *Computer-Integrated Manufacturing.* Industrial Press, New York, 1973.

[18] Tanner, W. R., *Industrial Robots: Fundamentals,* vol. 1, 2nd ed. Robotics International of SME, Dearborn, MI, 1981.

[19] Hannemann, J., *Robots: Selection and Installation.* Berguall Productions, Garden City, NY, 1986.

[20] Wallace, Susan, "Automation Cost Decisions By Team." *Managing Automation,* August 1988, p. 36.

[21] Knight, Robert, "Plan First, Then Use CIM." *Software News,* December 1987, p. 61.

9

The Human Factors
of CIM

Development Process

Changes take place to solve or to address problems. Before a problem can be solved, the system analyst must understand the problem, the characteristics of the organization, and its personnel practices. To develop a computer-integrated manufacturing (CIM) system for implementation in the company, the system analyst should follow a methodical approach to ensure that organizational needs are met. A major responsibility of the system analyst in the implementation of CIM is to conduct a feasibility study, perform a system analysis, develop a system design, and install the system. The analyst must perform a preliminary investigation to determine the feasibility of developing a system. If the findings are acceptable, a system analysis is conducted. Following the system analysis, a system is designed on the basis of the needs of the organization. Once the design has been completed, the CIM system is implemented. Figure 9.1 is an overview of the development phases of a CIM system.

In this chapter, the feasibility study, system analysis, and system design are presented. System implementation was introduced in chapter 8; this chapter gives a perspective on the development phases and the human factors involved in the initial development of the system. It is discussed in more detail in Chapter 10.

Feasibility Study

The purpose of a feasibility study is to determine if a problem exists and if so, to ascertain its nature and scope. The feasibility study can also be conducted to make modifications to an existing system that no longer meets the needs of the

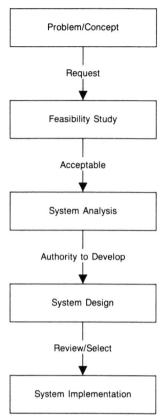

Figure 9.1 Overview of CIM Development Phases

company, or to ascertain if a new system is needed to solve a problem not previously identified. The feasibility study implies a study of the practicability of a proposed project and involves an analysis of the total requirements for a human, economic, technical, and environmental evaluation of a proposed solution. The study should give management the information necessary to decide whether the project should be attempted. The request for the study can come from management, from users, or from within the systems group. The feasibility study can also be referred to as a preliminary investigation.

A company's business ambition determines the CIM strategy. Therefore, a company must first analyze its business and determine a priority of business objectives before it can start to draw a plan for CIM.

Data are gathered from a variety of sources to investigate the problem. Data are usually obtained by conducting interviews, observing the way data are processed, studying documents, and administering questionnaires. The following should be accomplished during the feasibility study:

- The problem is clearly defined.
- The nature and scope of the problem are determined.

- The resources and time required to complete the project are estimated.
- The time it will take to modify or develop the system is estimated.
- Conclusions are formulated and recommendations are made.

The analyst making the study provides both a written and oral report to management. The recommendation from the findings of the study may be either accepted or rejected. The feasibility study can have one of three results: (1) The recommendation to do system analysis is accepted and scheduled, (2) management may ask for additional information, or (3) the project may be dropped or put on the shelf indefinitely.

System Analysis

The major objectives of system analysis are to further define the problem and to determine the best way of solving the problem. Before a solution can be formulated, a study of the current system should be done to ascertain how it is functioning. Characteristics of the current system can provide insight into solutions and can be incorporated into the new or modified system. Investigation of the current system can provide a framework within which to address the goals and objectives of the organization in the current environment.

Users, management, and the analysts must be directly involved in determining the objectives that spell out the need for a CIM system. They should consider constraints that may limit the scope of the project. The constraints may be of a financial, legal, physical facility, human resources, or economic nature.

The investigation should probe deep into individual problem areas, functions, and details of operations to gather quantitative data essential for the determination of requirements for the design of the system. Therefore, the primary purpose of studying the present system is to survey it and its operation to understand what the system does and what it should do. An overall picture needs to be obtained to understand the position of the company, the structure of the organization, and the relationship among various functions.

Tools used to gather data during system analysis include interviewing, observing, inspecting records and documents, and administering questionnaires. Interviews are a major source of data. Management and operational personnel are questioned on how and why tasks are being performed. Observations are another way to collect data. By observation, the analyst can determine the accuracy of standards manuals and the difference that exists between the formal and informal organizations. Observations are generally made to determine (1) work flow, (2) physical location and arrangement of work stations, (3) who does what task, and (4) how each task is performed.

Inspecting records and documents is a method of data collection used by analysts to determine the internal and external environment of the organization to determine who receives reports and the flow of information, along with determining the importance of the information. By inspecting books, periodicals, and brochures, analysts can gain insight into trends, alternatives, and product specifications. The questionnaire is a popular data collection method when a large number of people need to be questioned. Questionnaires are used to collect details about the volume of input and output of systems operations.

After all data are gathered, they must be analyzed and interpreted in terms of the objectives of the study. The decisions made after data are analyzed will be influenced by the constraints imposed by management. The results of the analysis are presented to management in both a written and an oral presentation. Recommendation from the study will cause management to do one of the following:

1. Approve the project as presented and schedule the next phase.
2. Approve the project in part but ask the analyst to modify it.
3. Shelve the project until some time in the future.
4. Drop the entire project.

System Design

System design involves developing alternative models that satisfy the goals and objectives set up in the system analysis report. System implementation results when a design is approved by management and incorporated into the organization.

In system analysis, the focus is on what is being done in the current system and what should be done according to the requirements stated in the analysis. In the design phase, the focus is on how a system can be developed to meet the requirements.

During the system design stage, the following steps are taken:

1. *Review Goals and Objectives.* The design should conform to the goals and objectives established in the analysis report. The design should reflect the needs of most of the users and the needs of the organization.

2. *Develop a System Model.* A model of the system's major components is established. The model includes the (a) system input, (b) system processing methods, (c) system output, (d) system database and files. The major interactions among the components should be shown in the model. Each component is then designed from the model.

3. *Evaluate Organizational Constraints.* The resources and constraints of the organization must be evaluated to design components that will function within the limitations. The constraints should be viewed in respect to their impact on the system.

4. *Develop Alternative Designs.* In most functions, there is more than one way to approach a problem. A system requires a great amount of interaction among subsystems. Therefore, there should be more than one design to select from. With alternative designs, the best of them can be integrated to develop the best system.

In the alternative systems, the feasibility, tentative input forms, output format, files, and clerical procedures should be included.

5. *Perform Feasibility Analysis.* The feasibility of all alternatives must be uppermost in the mind of the designer, who must consider the feasibility of all staff, equipment, laws and regulations, time, and cost.

6. *Perform Cost/Benefit Analysis.* Because organizations have limited financial resources, projects that promise the greatest return on costs of development are those that are implemented. Costs and benefits must be quantified. Benefits can sometimes outweigh the cost factor. It is not always necessary that a plan forecast positive economic benefits for it to be considered feasible.

7. *Prepare Design Report.* A design report should explain how the design will address the requirements for the CIM system. The system analysis results, the proposed design, the implementation requirements, and the alternative recommendations should be detailed in the report. Each aspect of the report should be clearly explained, using flowcharts as necessary. The design report is submitted to management. Management will (a) approve the recommendation, (b) approve recommendations with changes, (c) select none of the alternatives, or (d) do nothing.

System Implementation

The objective of the system implementation phase is to achieve a fully documented operational system. In this phase, computer-based programs, manual procedures, and control features are built and tested as the installation of the entire system takes place. This phase also includes the preparation of documentation that describes the features of the CIM system, its components, and how it is to operate.

Implementation of the system requires careful scheduling and precise handling of resources. It includes mechanical and human interfaces and backup procedures. Extensive training, testing, and control are required in the implementation phase to successfully install the system. Chapter 10 discusses the implementation phase in more detail, describing the components and procedures necessary to implement a successful system.

A system must be designed that can be used by people because the purpose of the system is to help people solve problems. If people do not use the system, it will not become operational. Therefore, the human factor in designing and implementing a system is crucial to its success.

The remainder of this chapter discusses human factors that should be considered to facilitate a successful system. Systems analysts and managers should not just concern themselves with the technology involved in the analysis design and implementation of a system but also the integration of human resources that can determine the success or failure of a system.

The Human Side

The success or failure of any system is dependent on the people who will use it. If they do not use the system, it will not be the productive tool it was intended to be. It is often said that people are the greatest resource for an organization, and indeed they are. However, all too often, the impact of the system on the employees is given little attention during development.

Other than economics and job security, the greatest personnel problem seems to be job dissatisfaction. This area has had great impact on system failure and success. Job dissatisfaction causes problems such as stress, absenteeism, low performance, and low morale.

In the early years of industrial organizations, personnel departments (if they existed at all) were at most a one-person operation making decisions on hiring, firing, input, and output. Today, organizational structures are more complex, having departments and managers for each aspect of the operation. Human resource departments serve line managers and advise them on human resource activities and control to get consistent treatment of employees. The functions of human resource departments have become broader to satisfy the needs of the employees and the organization.

In spite of concern for employees and the organization, there remain discontent and unproductivity. The discontent of workers has brought many internal and external influences to bear on the workforce: labor laws and regulations, labor unions, job analysis and redesign, performance measurement, personnel planning, valid employment predictors, and career management programs. If systems are going to succeed and work is to be satisfying, the needs of the employees must be addressed. The following must be considered:

- Employees' well-being
- A safe and healthy work environment
- A comfortable work environment

Employees' Well-Being

When a company provides for its employees' well-being, it considers their job satisfaction, the package of fringe benefits they have, retirement benefits, job security, morale, and cooperative spirit. Productivity is the end result of satisfying the well-being of employees. A satisfied worker is a productive worker. Great skill and knowledge are required to operate today's increasingly sophisticated equipment, and employees expect to be able to use their skill and knowledge on the job to gain self-esteem and self-actualization.

Challenge and opportunity are important to employees in an electronic environment. Expectations have become stimulated, and employees are less willing to tolerate work that is not meaningful to them. Loyalty appears to be to oneself first, then to one's profession, and then to the employer [3].

Employee stress can result from an environment that combines high technology and people. Increased noise levels, poor lighting, inappropriate furniture, or introduction of new equipment can affect employee stress. In an era of complexity and rapid change, stress is a problem for individuals and for organizations, and seems to be a special problem for those in a data processing environment.

Warrick, Gardner, Couger, and Zawacki [16] describe stress as a person's mental, emotional, physical, and behavioral response to anxiety-producing events

[16]. They point out that mismanaged stress can decrease performance, motivation, and alertness, whereas managed stress can increase alertness, sensitivity, and effectiveness. In addition, they indicate that what is less obvious about stress is the multiple effect that one person's coping response can have on others in a department. Some individuals may deal with their stress by stressing others.

What does this information mean for managers? What are the implications? Warrick, Gardner, Couger, and Zawacki point out the following from their study: "Several symptoms may indicate that the staff is overstressed. Watch for patterns of fatigue, irritability, increase in mistakes, missed deadlines, and drops in productivity. Symptoms might also include low morale, and high turnover" [16]. Their research indicates that managers should learn to understand and manage a stressful environment by doing the following:

- Educating themselves about the consequences of stress.
- Identifying personal and job stressors and their effects.
- Selecting those stressors they can do something about and reduce, eliminate, or manage them.
- Developing a plan for coping with stress.

A Safe and Healthy Work Environment

Of itself, an electronic environment such as that found in computer and communications industries does not pose health hazards, but the way the technology is used can affect employees. A company should consider certain health hazards that can be present.

Video Display Terminals

The prolonged viewing of a VDT can cause problems. It can affect an individual's ability to focus from far to near and vice versa. According to OSHA (Occupational Safety and Health Act) four hours of uninterrupted VDT work can impair the ability of the eyes to focus for up to 30 minutes. The institute recommends a 15-minute rest break after two hours of moderate work on the VDT. If the work on the VDT is intensive, the rest breaks are recommended after one hour.

The American Optometric Association has found a 20 percent increase in eye strain among VDT workers. Therefore, Franche and Kaplan suggest longer rest breaks for VDT workers [2]. They also recommend that workstations be arranged so that the VDT worker can look up and out across the room occasionally.

VDT Radiation

In 1981, NIOSH (National Institute for Occupational Safety and Health) published discussions about radiation as a possible concern to users of VDT. They link regular use of the VDT to eye strain, headaches, nausea, lower and upper back pain, and stress [1].

Machine-Paced Work

Other NIOSH findings are that machine-paced work can cause stress among workers—for example, work on an assembly line. Workers should be able to keep the pace of their work at a reasonable level.

Ergonomics

Ergonomics is the science addressing the compatibility of people and machines. Designing the model electronic environment is an increasingly important task that affects employee morale, productivity, and job satisfaction. The successful design of a system requires that the interface be designed to serve the needs of the worker in both the organizational and personal content. The following must be considered [3]:

- Access floors
- Power systems
- Wall partitions
- Ambient lighting
- HVAC system (heating, ventilating, and air conditioning)
- Acoustics
- Buffer zones
- Workstation components

Human Resource Planning

Motivation

If half the potential of human resources could be used, the productivity increase would exceed that possible from technology. What inner mechanism motivates people to give their all in certain situations but not in others? Why do people behave the way they do? Is it worth the effort of management to explore answers to these questions? Social scientists have developed hypotheses about behavior that can help managers. The hypotheses are differentiation, needs models of motivation, and expectancy models of motivation.

Differentiation

Behavior differences between individuals are produced by physical differences, mental capabilities, life experiences, culture, and perception of a situation. For example, with age often comes a reinforcement of experiences and interpretations that may lead to resistance to new methods and concepts.

Needs Models of Motivation

The basic physiological needs are usually satisfied through the medium of money. According to the need models, we might expect that constant increases in money decrease in importance once the physiological needs are satisfied. Money remains important only to the extent that it contributes to social or the fulfillment of ego needs.

Another example of the needs model is the case of people to whom security is a high-priority need. We could place them in stable organizations, provide security against layoffs and firing, offer steady growth, and provide comfortable medical, health, and pension benefits to meet their motivational needs.

Some motivational needs are not easy for management to handle. For example, the need for the esteem of others can be a very strong motivation, but self-esteem may be fulfilled in many ways, making it difficult for management to develop incentives based on the need. Self-esteem is essentially a matching of a person's performance with his or her values and of those values held by the individuals in the organization.

Highest in the hierarchy of needs is self-fulfillment, which is the need to achieve all that the individual is capable of, and to achieve it in a kind of work that the individual enjoys. Modern managers are now taking this need into account. Job enrichment through vertical and horizontal expansion of duties, increased participation of individuals in setting their own goals and in decisionmaking, and a more careful search for the interests and capabilities of each employee are being carried out by progressive companies.

Expectancy Models of Motivation

In the expectancy model, motivation is viewed as being dependent on the strength of an individual's desire for a set of goals and the likelihood that a specific type of behavior will lead to their achievement.

Despite the differences among the theories of motivation, the common message for managers is: Determine what each worker wants, develop the tasks, and present an environment that will permit the worker to achieve his or her goals. A reward system can be developed for achievement of goals that match organizational goals.

The role of managers in implementing motivating systems is vital. Only managers can balance the human and technical aspects of systems. They need to ensure that their employees' jobs include diversity, flexibility, variety, and autonomy. These needs do not change when employees use a computer to perform their work [9].

Effect of Technology on Behavior

A barrier to the acceptance of technology is that individuals do not understand it or refuse to accept the reasons for changes.

People are the essential ingredient in all aspects of the organization. They are the factor that determines the survival of the company and its objectives. Therefore, effective management of people is a major consideration in planning overall organizational strategy. Employee reaction to the installation of a new system can range from failure to use the output generated by the system or deliberate sabotage of the system.

Barriers to Change

People resist change because it upsets their established patterns. They are also threatened by a possible job loss. The barriers individuals erect against change can be classified as perceptual, emotional, or cultural.

Perceptual barriers. The individual may view a change as a personal criticism. To avoid this reaction, explain the reasons for the change before it is put into effect. The worker may not be willing to retrain or relearn skills and knowledge.

Emotional barriers. The individual may have apprehension or biases against changes, especially those brought about by new technology, or caused by their personal prejudices.

Cultural barriers. Individual social relationships develop both on and off the job. Employees often do not want social relationships altered by what they consider to be outside influences (the new technology). They also believe no contributions are made to developments in the organization and nothing is gained by working alone as an individual.

A change disrupts a person's frame of reference if it presents a future environment where past experiences do not hold. Such a change is likely to be viewed as a crisis. Many people try to cope by trying to maintain control over the situation. In the case of a new system, they may deny the change. They may also distort information they hear about it, or they may try to convince themselves that the computers will not change the status quo.

The project team's goal, therefore, is to build commitment to the new system to keep employees' support through the change. The greater the change to the current way of operating, the greater the commitment needs to be from the top executives, project development team, and expected users of the system.

Group Dynamics for Information Systems

Organizations are made up of people, not boxes on organization charts. Therefore, managers and system designers must not isolate themselves from the actual organizational dynamics. Management should use information about groups in the design of systems and should acknowledge and strengthen groups in the organization's environment.

Group dynamics designates the forces and behavior that occur within a group. The linking of groups and subgroups of a business organization form the information organization.

Six types of informal groups have been identified by sociologists, as follows:

1. The *total organization* consists of all the many interlocking groups or subsystems in the entire organization.
2. *Large groups* form over some issue of internal politics. For example, nonunion groups versus union groups, young executives versus old-time executives.
3. *Primary cliques* form when workers are located together for work or when employees have similar job and common interests. For example, whether maintenance crew or top executives, people who work together often take their meals together.
4. A *clique* is a small group formed to gain some special power or social advantages. An example is a sports league competing within the organization.
5. *Friendship and kinship groups* form when generation after generation of the same families become employees. Friendship groups form because of close social and neighborhood ties.
6. *Isolates* are individuals who are loners and do not attach themselves to any group or shift from group to group.

Without proper consideration of the behavior of people in the business organization setting, the best technically designed system is likely to fail because a new system represents a threat to individuals' organizational relationships and psychological needs. Therefore, management must do more than involve everyone in the development of the system; it must also consider the newly emerging organization pattern and the groups within it.

Resistance to CIM

The reluctance of factory managers in the United States to apply CIM appears to be centered on risks associated with the technology. There is an additional, and perhaps larger, risk if advanced manufacturing technology such as CIM is adopted without the right human resource strategy for an organization. One of the greatest risks is attempting to move toward CIM without thinking through the changes in corporate culture that have to be made. If management does not conduct a social revolution at the time CIM is implemented, the technology will lower the rate of return and the organization will fall behind in the competitive market.

Effecting Change Without Resistance

Three positive steps for effecting change that are based on our knowledge of organizational behavior are the following:

1. Create a climate for change.
2. Develop effective agents of change.
3. Modify the required organizational system in the light of anticipated emergent behavior.

Create a climate for change.   A climate for change may be obtained by getting managers and workers to feel dissatisfied with the present system. Discussions can be focused on what is wrong with the present system and ways for revising it. The participants will be left with a feeling that changes are needed.

Develop effective agents of change.   Within any organization there are informal leaders and technical leaders to whom other members of the group look for protection and security. Management and the system designer must identify such leaders and win their support.

Modify the required organization.   As the technical side of the MIS is developed, the organizational requirements must be made as anticipated emergent organizational behavior dictates.

Management, Information, and the Systems Approach

There are three important tasks of management and hence goals of the organization:

1. Match the capabilities of the institution (business enterprise, university, public agency, etc.) to various needs of the environment and select specific missions from among these opportunities.
2. Establish a work environment and allocate resources for maximum productivity of the total system.
3. Manage responsibilities to and the impact on stockholders.

It is evident that the first and second steps in the planning process depend heavily on the availability and use of critical information. It is hard to imagine the manager trying to develop any of the three major types of plans without first knowing planning premises that will permit adequate evaluation of alternative courses of action to achieve the plan.

Productivity

Two basic business trends have an impact on manufacturing: (1) Direct manufacturing jobs are becoming more mechanized, and (2) those tasks not mechanized are being or will be done overseas where labor is cheap [4]. Manufacturing is going the way of agriculture; 3 percent of the agricultural labor force produces 120 percent of the nation's requirements. Automation increases the ability of an individual worker to control more output. The trend toward automation may create more jobs of monitoring rather than doing. This situation can be incompatible with the human trends of education and expectation. Therefore, technology must be considered an equal partner in the manufacturing plant of the future.

Merely installing new technology is not the answer to making information systems pay off. According to Blair and Uhlig, companies blunder if they stress technology and ignore the demands of employees to have a voice in the design of the systems [13]. Most systems do not work behaviorally. Blair and Uhlig believe that systems demotivate rather than motivate the employees who use them. They suggest two approaches to solve this problem:

1. Add occupational development persons and social scientists to the system design teams.
2. Redesign methods and procedures to gain efficiency and effectiveness from the technology.

These approaches help to improve the human side of work [13].

CIM Education Tasks

The increasing number of people using CIM creates a tremendous training and education task for the organization [6]. Education is needed throughout the organization. Top executives reviewing organizational plans for computer and communications systems need to have enough technical expertise to make intelligent decisions. Middle managers need to learn how to manage in the new technological environment because they will have the responsibility for making sure the use of the system is successful. End users need to learn CIM concepts and how to use specific computers, applications, and on-line services. Therefore, the education of employees across the organization becomes essential as users and their managers assume more responsibility for the funding, justification, development, and use of CIM. The systems department plays a coordinating, supportive, and standards-setting role as part of its responsibility to build and maintain the technical infrastructure for the system.

In the three sections that follow, the areas listed below on organizational education needed for a CIM environment are discussed in more detail.

1. Executive education
2. Employees education
3. Users education

Executive Education

Top-level policy-setting executives must understand the nature and ramifications of information technology to make the proper decisions for their organizations [6]. A good understanding of CIM technology, derived from a coordinated education program, will give executives the ability to make the informed decisions necessary to guide their companies in today's technology-driven world. Executives knowledgeable about computer and communication technologies will be able to make wise decisions when allocating resources, planning, setting technological direction, and evaluating.

For executives to achieve this expertise, both education and training must be dealt with. Education concerns concepts and understanding, and training emphasizes skills. Most user training focuses on the skills required for nontechnical users to interact with a computer directly. To use a computer competently, however, users need to understand basic data processing concepts, and, therefore, user training should involve both education and training.

Systems executives play an important role in the successful implementation of CIM. They can nurture or destroy a new system, consciously or unconsciously, by their support or resistance. Therefore, they need to become familiar with the risks and problems of new systems in their organization. They must make sure that their subordinates' jobs, health, and safety needs are met by the new system as part of managing the transition from the old to the new.

Types of Executive Education Programs

A wide range of educational programs can be offered to the executive and can be categorized into three areas as follows:

- Informal education program
- Semiformal education program
- Formal education programs

Informal education programs involve the executive in previously established education channels, which include reading publications, sharing knowledge through subordinates, and individually viewing demonstrations.

Semiformal education programs provide voluntary courses to introduce new subjects and to update knowledge. They include executive briefings, brown-bag lectures (videotape courses given during the lunch hour), and short seminars.

Formal programs provide intensive one- to three-day courses at which attendance may be mandatory. Two variations of a formal program are a single one- to three-day session or a series of short sessions.

In many new systems, little thought is given to the cost of training and education. Many systems people consider these areas to be secondary in importance, considering technical features to be of primary importance. However, education and training are two of the most important criteria for success. A technically successful system is an operational failure if people will not use it. As CIM becomes the heart of the organization, systems success or failure can mean business success or failure.

Employee Involvement

Success of a CIM ultimately depends on the individual behavior of the organization's members, who include users, managers, and systems specialists. Behavior evolves from individuals' predisposition to engage in their assigned tasks. Their commitment to the tasks is strongest when they believe it will be worth it to them to contribute their time and effort.

According to Zmud, the impact of information systems (IS) is believed to be derived from a variety of factors [7]:

- Formal education or training programs
- Attitudes about purpose of IS and need for change
- Implementation experience

These factors are all influenced by the climate of the organization's information system. The climate is governed by how these factors are handled by management in addressing the needs of employees and the organization. Therefore, establishing an overall organizational climate conducive to CIM's success is important and cannot be overemphasized.

The steps in establishing and maintaining a good organizational climate include having well-defined procedures for CIM, a clear understanding of the organization's function, enforcement of the relationship between the information function and the users, a structure for handling CIM responsibilities, and encouraging members to actively participate in the systems activities. A negative climate can impede individual contributions, whereas a positive climate permits and enhances individual contributions.

Factors affecting CIM behaviors include organizational size, environmental complexity, change, organizational structure, time frame, and the availability of resources [8].

Specialized skill, knowledge, and experience are needed to successfully develop information technologies. When information that reflects unique organizational characteristics is applied to CIM systems, organizational understanding becomes as important as technological understanding, which means that the need for in-house capabilities intensifies. Substantial investment in human resources is required to implement those applications producing the largest organizational benefits.

It becomes very important to have adequate user participation in providing for the organization's information resources. Sometimes it can be tempting for systems specialists to let the user relinquish responsibility for the information resources. Most systems specialists do not enjoy interacting with users because it distracts the users from technical tasks that they would rather do.

Systems users and specialists tend to differ in background and interests, which can result in communication difficulties that inhibit the cross-education that is so critical to successful implementation of information systems. These differences are reflected in approaches to problem solving, attitudes toward changes, and views of each other. However, the more users and specialists interact, the more they can learn about and from each other.

Implications of failure can be great. Failure of CIM tends to build because of the interrelated nature of systems applications and the impact on the systems climate. Failure represents a high cost that results because project benefits are not realigned, the existing system is retained, there is misuse or underuse of the system, personnel are not interested in the information resource, or there is distrust of the information function.

One of the more common obstacles to the successful implementation of CIM is that systems are often conceived by an official or committee at corporate headquarters and then pushed down to plant management. This is almost a sure-fire formula for failure. The top people have to understand user needs and opportunities and must get the user involved in the system specifications.

Job Redesign

Manufacturing jobs should be designed to match the education and expectation of the workplace. The trend in factories is the higher educated worker who needs more than security. This worker needs a match in job design.

The study by the American Productivity Center in Houston, Texas, in 1983 addressed three main aspects of productivity improvement: human resource development, automation technology, and environmental considerations. The largest gain in productivity occurred when all three aspects were addressed and integrated into solutions. However, the findings of the study indicated there were very few cases where all aspects were considered by management. The message evident in the study is that employees want to be involved in the decisions when their working environment changes [5]. They want to be productive providing they have the right tools and are a part of the decisionmaking process of their job design.

Taking into account goals, current attitude problems, and options for using technology, job redesign can be conducted by employees in small groups. Outside help can be provided to demonstrate various ways in which jobs can be reconstructed. The job redesign activities should decide which technological options will be used and how they will be used [6].

Many of today's jobs do not encourage employees to experiment with new ways of working. If companies want to take full advantage of new information systems, they need to encourage employees to discover new ways to use these systems. More employee participation in job design can and will have far-reaching effects— such as altering methods of authority, means of training, places of work, organizational structure, and other aspects of work that many employees take for granted.

Meeting User Needs

A number of factors contribute to the success of a system. A crucial factor is that the system meet the requirements of those who will use it in their day-to-day duties.

In the past, users have sometimes been unable or unwilling to articulate their needs. Designers have sometimes produced systems that have been more oriented to using the latest technical development or to serving the designer's ideas of managers' requirements than to meeting the existing needs of the organization.

The emphasis has also often been on producing the system in a shorter period rather than allowing the necessary time to study and develop a system that meets the requirements of the user. The objective of the installation of a system must be to provide each user with the type of information that will be used most effectively in circumstances confronting the individuals involved [12].

Specifying and implementing the requirements of a system is best undertaken with cooperation from both managers and specialists. The cooperation brings about the design and the implementation that incorporates a sensitivity to user requirements and to human factors. In this way, the system becomes an adjunct to managers' work rather than technology that has been imposed on them.

Employing experienced staff in the design and implementation of CIM is a major key to meeting the prescribed requirements of the system. Sometimes there is reluctance from top management to assign the best individuals to the development task or to allow enough time to allow them to make a contribution. There may also be the temptation to use less experienced systems specialists because they are less expensive or because there is a shortage of more experienced persons.

Potentials and Limitations

According to Pana the business environment is more turbulent because of the rate of change and complexity [10]. Some companies cannot respond quickly to change because their structure requires communication up and down the chain of command, which can take a lot of time. Pana believes there should be less emphasis on strict hierarchies and more emphasis on dispersed decisionmaking. His premise indicates that hierarchies work best in a stable environment but not well in a dynamic one.

A recognized benefit of new information technologies is the greater means of communication options. Store-and-forward computer, voice message systems, and teleconferencing systems permit groups of people to communicate more easily. Consequently, people can get a broader picture of the operation, and they can coordinate and control work more easily [11].

It is necessary to develop a clear idea of the benefits that will accrue from the system at an early stage in its design and implementation. Many of the benefits of an information systems are not easily expressed in tangible terms, however.

In the early days of information systems, systems were developed to service transactions and routine decisions at the lower and middle levels of an organization. The objective of the system was to reduce cost or to change a labor-intensive operation to an increased capital operation. The benefits of an increased capital operation were clearly measurable in terms of cost reduction or maintenance service per unit of human effort. The expected benefits were not always achieved because estimated techniques are not always reliable.

In the human area, benefits from the information system can be manifest in greater awareness and alertness of the members of the organization, which can result in the following:

• Clearer appreciation of the organization's objectives, directions, values, and preferences.

- Increased understanding of the organization's operating environment with more awareness of the social, technological, and natural elements of that environment.
- Greater appreciation by managers of the progress toward achieving the goals and objectives of the organization and the resources necessary to continue the progress.
- Increased communication and exchange of information among members and divisions of the organization.
- Greater availability of historic data and information for use in planning and related activities.
- Avoidance of the cost of not implementing a comprehensive information system.

It would be ideal to be able to quantify the intangible benefits of the implementation of an information system. However, progress in that direction is very slow, and managers will have to continue with a completely quantified and well-defined method of estimating the benefits that may accrue.

Financial Justification

The implementation of CIM is expensive. The expense is not easy to justify by using traditional financial yardsticks. Few, if any, CIM projects would satisfy return-on-investment criteria. A CIM project with a seven-year payback period could give a better solution than one with an expected two-year payback. Intangible factors normally not taken into account have now been incorporated into financial justifications for CIM systems.

Because investment in CIM can be expensive, management must consider the investment as essential to achieve the company's long-term goals and objectives. A long-term commitment to the investment must be made, and CIM funding should be part of future planning because CIM is evolutionary.

In his book, *Information Payoff: The Transformation of Work in the Electronic Age,* Strassman discusses the areas to which executives should direct their energies in order to make investment in information technology pay off [14]. He addresses these areas with the following questions to managers:

- Do you know the total cost of your investments?
- Do you know the organizational consequences of your investments?
- How do you measure the effects of your investments?

Total Cost

Total cost is the management of labor costs that determines whether the investment pays off. The total cost in financial justification is based on several factors. Some of these factors are discussed below. The management of labor costs should be given high priority. It contributes most in determining whether the investment pays off. One must not only look at the technical cost of the system but also at the cost of people. In 1985 Strassman estimated that companies would need to spend at least $5,000 (and perhaps as much as $20,000) per employee to make environmental changes [14]. In addition, organizational costs must be considered in the total cost. Organizational costs include training costs (both initial and on-the-job training); start-up costs, which include such costs as the beginner's lower productivity; and the costs of external staff support such as consultants, auditors, and temporary assistants.

Organizational Consequences

Benefits of the technology come from improving group communications rather than improving individual tasks. Reducing the number of internal communications and increasing the number of external communications to customers, suppliers, and others increases an organization's productivity. Strassman believes managers should plan how the organization should change to achieve the business objectives; then they should select the technologies to achieve these goals. To make the system economically viable, they should concentrate on training, improving intragroup communications, and trimming work flows.

Measurement of Effects

The most common methods for evaluating changes in productivity is by distributing work into small segments. If each segment is more efficient, then productivity is said to have increased.

In discussion of cost justification, you need to be familiar with the term *value added,* which is used to justify automated systems for professional or management workers. The addition of automated systems assumes improvement in performance or productivity. Improvement is measured in *soft dollars. Soft dollars* are those that a company can recapture on the basis of increased productivity, and help pay for the CIM system over a long period [15].

The costs of CIM can be categorized as either one-time costs or recurring costs. One-time costs can include site preparation, freight, salaries, and other direct expenses associated with implementation and conversion activities. Recurring costs include improved public relations; accurate, consistent data; improved product quality; and direct labor cost applied to a specific task or job.

Benefits of CIM can be categorized as follows:

- Cost reduction. Conversion to the system achieves cost-related benefits.
- Cost avoidance. Conversion to the system negates commitments of other organizational resources.
- Value enhancement. Conversion to the system provides improvement capability that would not otherwise happen.

Cost controlling leads to cost reduction. The initial cost controlling efforts are directed toward determining what the costs are, what they should be, and how best to maintain them at satisfactory levels. These efforts seek to reduce costs by improved methods, procedures, and cost standards.

Cost reduction should not be considered to be only a management job. Every employee should participate. Employees should be interested in reducing costs because it contributes to employee security. Cost reduction contributes to keeping an organization fit to continue to operate successfully and to meet its responsibilities.

Exercises

Homework Assignments

1. Explain the significance of a feasibility study as part of developing a system.

2. Discuss the difference between system analysis and system design.

3. Explain and discuss two items that should be considered in addressing the human side of computer systems.

4. Using one the motivation models discussed in this chapter, develop a hypothetical situation and apply one of the models to the situation.

5. List and discuss behavior barriers to change.

6. Explain the significance of considering group dynamics in developing a system and also discuss the characteristics of various informal groups.

7. List and discuss methods for effecting change in an organization without resistance.

8. Discuss the impact of technology on training in the organization.

9. How important is executive training in information technology, and what types of executive training programs can be offered?

10. Explain how employee involvement in system development may affect the success or failure of a system.

11. Discuss the potentials and limitations of information technologies in organizations.

Self-Study Test

1. Changes take place in organizations just for the sake of change.
   True   False

2. The feasibility study is designed to determine if a problem exists and to ascertain its nature and scope if one does exist.
   True   False

3. Another term for a feasibility study is a preliminary investigation.
   True   False

4. Management, users, and analysts should not be directly involved in determining the objectives for the need for CIM.
   True   False

5. The primary purpose of studying the present system is to survey the current system and operation in order to understand what the system does and what it should do.
   True   False

6. By inspecting books, periodicals, and brochures, insight can be gained regarding trends, alternatives, and product specifications.
   True   False

7. The results of the systems analysis are not presented to management in a written or oral presentation.
   True   False

8. In the system design phase, the focus is on studying the current system.
   True   False

9. The purpose of the system implementation phase is to achieve a fully documented operational system.
   True   False

10. Success or failure of a system is dependent on the people who will use it.
    True   False

11. Machine-paced work does not cause stress among workers.
    True   False

12. Behavior differences between individuals are produced by physical differences, mental capabilities, life experiences, culture, and perception of a situation.
    True   False

13. According to the needs model of motivation, money remains important only to the extent that it contributes to the fulfillment of social or ego needs.
    True   False

14. If half the potential of human resources could be used the productivity increase would exceed that possible from technology.
    True   False

15. The highest need in the needs hierarchy is the need for self-fulfillment.
    True   False

16. On the whole, managers do not play a vital role in implementing systems.
    True   False

17. The feasibility study implies the practicability of a proposed project and

involves an analysis of the total requirements for a _____ evaluation of a proposed solution.

1. Human
2. Economic
3. Technical
4. Environment
5. All these.

18. A feasibility study can result in

1. A recommendation to do the system analysis.
2. Management's asking for additional information.
3. The project's being dropped or put on the shelf.
4. Items 1 and 2 only.
5. Items 1, 2, and 3.

19. Constraint(s) that may limit the scope of a CIM development project may be

1. Financial
2. Legal
3. Human resources
4. Economic
5. All these.

20. _____ are a major source of data collections.

1. Interviews
2. Observations
3. Document and record inspections
4. Questionnaires
5. None of these.

21. Observations are generally made to determine

1. Work flow
2. Physical location of work stations
3. Who does what task
4. How each task is performed
5. All these.

22. In the _____ phase the following steps are token review objectives: develop a model, evaluate constraints, develop alternative models, conduct feasibility analysis, conduct cost/benefit analysis, and prepare recommendation report.

1. Analysis

2. Design

3. Feasibility

4. Implementation

5. Review

23. In addressing the human side of computer systems, _____ should be considered.

1. Providing for employees' well-being

2. Providing a safe and healthy work environment

3. Providing a comfortable work environment

4. Items 1 and 2 only.

5. All these.

24. _____ is your mental, emotional, physical, and behavioral response to anxiety-producing events.

1. Technology

2. Stress

3. Environment

4. Information

5. Sensitivity

25. _____ is the science addressing the compatibility of people and machines.

1. Ergonomics

2. NIOSH

3. VDT

4. All these.

5. None of these.

26. _____ views motivation as dependent on the strength of an individual's desire for a set of goals and the likelihood that a specific type of behavior will lead to the achievement of the individual's goals.

1. Differentiation model of motivation

2. Needs model of motivation

3. Expectancy model of motivation

4. None of these.

5. All these.

27. _____ designates the forces and behavior that occur within a group.

1. Information systems

2. Barriers

3. Change

4. Group dynamics

5. None of these.

28. A _____ group formed to gain some special power or social advantages is considered a

1. Large

2. Clique

3. Kinship

4. Primary

5. All these.

29. The two facets of change are

1. Technological and social

2. Behavioral and psychological

3. Familiar and unfamiliar

4. Good and bad

5. None of these.

30. The term used to justify automated systems for professional or management workers is

1. Fail soft

2. Soft dollars

3. Value added

4. All these.

5. None of these.

ANSWERS TO SELF-STUDY TEST

1.	False	**7.**	False	**13.**	True	**19.**	5	**25.**	1
2.	True	**8.**	False	**14.**	True	**20.**	1	**26.**	3
3.	True	**9.**	True	**15.**	True	**21.**	5	**27.**	4
4.	False	**10.**	True	**16.**	False	**22.**	2	**28.**	2
5.	True	**11.**	False	**17.**	5	**23.**	5	**29.**	1
6.	True	**12.**	True	**18.**	5	**24.**	2	**30.**	3

References

[1] "Potential Health Hazards of VDT." NIOSH Clearing House on Occupational Safety and Health (4676 Columbia Parkway, Cincinnati, OH 45226), Fall 1981.

[2] "Race Against Time: Automation of Office Work 9 to 5." (1224 Huron Road, Cleveland, OH 44115) April 1980, pp. 31.

[3] Stallard, John J., and Terry, George R., *Office System Management.* Irwin, Homewood, IL, 1984, pp. 321–324.

[4] Cross, Kelvin F., "The Factory of the Future Depends on Successful Integration of Automation and Job Design." *Industrial Engineering,* Vol. 10, No. 1, January 1986, pp. 14–18.

[5] American Productivity Center, "White Collar Productivity: The National Challenge." 1983.

[6] Sprague, Ralph H., and McNurley, Barbara C., *Information Systems Management in Practice.* Prentice-Hall, Englewood Cliffs, NJ, 1986.

[7] Zmud, Robert, *Information Systems in Organizations.* Scott, Foresman, Glenview, IL, 1983

[8] Ein-Dor, P., and Segen, E., "Organizational Content and The Success of Management Information Systems." *Management Science,* 24, 1978, pp. 1064–77.

[9] "Manage the Impact of Systems on People." *EDP Analyzer,* Vol. 23, No. 5, May, 1985.

[10] Pana, C., "Microelectronics and the Design of Organizations." Working Paper No. HBS 82–67, April 1982, Harvard Business School, Cambridge, MA, Division of Research.

[11] "How Work Will Change: User Experiences." *EDP Analyzer,* Vol. 21, No. 4, April 1983.

[12] Mason, R.O., and Metroff, I.I., "A Program of Research on Management Information Systems." *Management Science,* Vol. 19, January 1973, p. 478.

[13] Blair, J., and Uhlig, R.D., *The Office Of The Future: Computers and Communication.* North-Holland Publishing Company, New York, 1979.

[14] Strassman, P.A., *Information Payoff: The Transformation of Work in the Electronic Age.* Free Press, New York, 1985.

[15] Dologile, N.G., *Using Small Business Computers.* Prentice-Hall, NJ, 1985, pp. 113–114.

[16] Warrick, A. A., Gardner, D. G., Couger, J. D., and Zawacki, R. A., "Stress." *Datamation,* April 15, 1985, pp. 88–92.

10

Implementing
a CIM System

After the decision has been made to develop a new computer-integrated manufacturing (CIM) system or to modify an existing system, a development team formulates a design to address the needs of the organization. Following the formulation of the design, the team is ready to implement the system. Implementation is the focus of this chapter.

Figure 10.1 illustrates the cycle of system development. System development can be viewed as the umbrella of the entire development process. The cycle starts with a concept to meet the needs of the organization. The concept is analyzed in relationship to the current system and the need requirements of the system users in the organization. After the concept is formulated in relationship to the current system, designs are formulated to develop solutions to the development of the concept. Models and prototypes are generated as part of the design process. The final step, which is the focus of this chapter, is to implement the design model that has been selected for the system. Once the design has been implemented, the system is reviewed, and if necessary, the cycle is repeated. Keep in mind in implementing a CIM system that there is no standard formula as to how a company should adopt computers to create a CIM plant. Companies operate in different ways and in different marketplaces, and each must devise its own strategy for success. Therefore, we can view the material presented only as suggestions for consideration for organizational implementation.

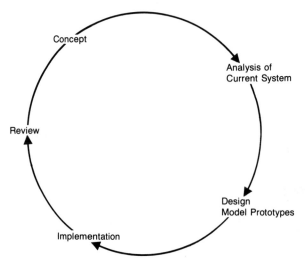

Figure 10.1 System Development Cycle

Implementation

Implementation may be viewed as putting product, service, or system into operation by means of the integration of the involved components. Implementation provides for the preparation of people and the environment for the development or modification of an existing or new system with the focus on meeting the organization's goals and objectives for CIM. The implementation process represents a conceptual framework that enables a company to accomplish the integration of product, service, or system. This framework also provides a systematic prescription for putting the design specifications into operation.

The investment costs of CIM can amount to thousands of dollars, and the success or failure of the integrated systems may determine whether the organization becomes or remains competitive in the industrial market. Therefore, it is critical that the best possible CIM system be designed and that a well-thought-out implementation plan be developed and executed.

Implementation Components

As mentioned in earlier chapters, CIM can link the islands of automation through enabling factors such as CAD, CAM, CAT, manufacturing planning and control, process technologies, robotics, and automated materials handling. Some specific strategies and key considerations in a successful implementation are the following:

- Planning
- Acquisition

- Testing
- Training
- Installation
- Security and control
- Maintenance
- Documentation
- Evaluation

Planning

Planning is the first step in the implementation process. It involves the development of plans, procedures, and schedules for the training, testing, acquisition, installation, evaluation, maintenance, security, control, and documentation of a CIM system. Furthermore, the purpose of implementation planning is to establish the tasks necessary to build a total plan to meet the goals of the installation of the system. Planning provides an organized procedure of assigning performance responsibility to specific individuals or departments and to estimate the duration of tasks.

Depending on the size of the project, the order of performance tasks may be described in a textual or graphic format. The larger the project, the more appropriate it is to graph a planning format.

Gantt charts, or network diagrams, are graphic formats that make visualization of the plan and schedules more clear. Specifically, the Gantt chart is used to document a schedule for each of the major tasks in systems development. It is a graphic illustration of task deadlines and milestones. An example of a Gantt chart is shown in Figure 10.2.

In addition to the involvement and support of top management, a full-time project manager and a training coordinator are required. If these positions are per-

Figure 10.2 Gantt Chart

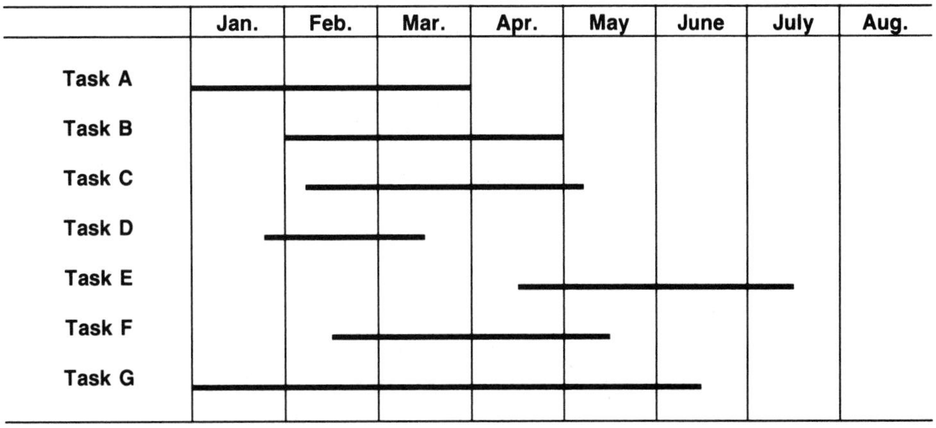

	Jan.	Feb.	Mar.	Apr.	May	June	July	Aug.
Task A								
Task B								
Task C								
Task D								
Task E								
Task F								
Task G								

ceived by management as not being important, it is a sign that management is not committed to implementing the system and handling its development. Commitment should be continuous in CIM. A system generally fails because of a lack of understanding of the purpose of the system, a lack of commitment, and a lack of understanding of behavioral changes that will take place.

Acquisition

The acquisition process is an evaluation of the proposals of manufacturers and other suppliers who furnish the hardware and software components required by the system specifications. Management must choose from among many different models and suppliers of hardware and software components.

Whether an organization is acquiring hardware, software, or consultant resources, the following sequence of activities should be followed:

- Determine requirements
- Prepare proposals
- Solicit bid
- Evaluate bids
- Negotiate contracts

Determine Requirements

The first activity, determining requirements, indicates the importance of the acquisition process. In the determination, careful analysis of the purpose behind the acquisition effort must be considered. An analysis must be made of the organizational activities supported, the objectives sought, and the means by which the objectives will be attained. Only after this analysis process is completed can hardware or software performance characteristics be realized.

Vendors need to know what a company expects over a period so that a schedule can be planned. The company should outline the expected monthly requirements from each vendor over an 18-month period. The company and the vendor must agree on the degree of variation from the schedule that can be allowed, and they must agree on penalties for orders that are reduced or increased within a stated number of days before a scheduled delivery date.

Preparation of the Proposal

The proposal is a formal document that defines why the resources are being sought, lists all specifications of the resources, and explains how the final selection of a supplier (vendor) will be determined. From the proposal, suppliers should be able to evaluate how their products or services meet the requirements.

Soliciting Bids

When the bid is solicited, the CIM design should be made available to the vendors who have been chosen. The CIM configuration may integrate hardware from more than one vendor. Among the means by which an organization can locate vendors are the telephone book, industry associations, other companies in the industry, professional groups, trade periodicals, and consultants. Each vendor selected should be provided with a request for proposal (RFP), which provides a summary of the criteria for the supplies being sought.

Evaluating Bids

Evaluating bids includes efforts to ensure that claims made by vendors are valid. Whatever the claims of hardware and software suppliers, the products must be demonstrated and evaluated. Five methods are generally used to assess vendor claims for their products: trial periods, references, benchmark testing, simulation modeling, and scoring.

First, hardware and software should be demonstrated and evaluated either on the premises of the user or by visiting the operations of other computer users who have similar hardware and software. Vendors sometimes allow potential clients to try out their products, but this method should be investigated carefully because some cost for transportation and installation may be incurred that would make direct evaluation inappropriate.

Second, other users are frequently the best source of the information needed to evaluate the claims of suppliers. A vendor should be willing to provide the names of such users. It is probable that the names of only satisfied clients will be made available, but even satisfied customers can experience problems and can provide useful information about both product and vendor.

Third, large companies that use computers extensively frequently evaluate proposed hardware and software by requiring that the products meet benchmark test programs and test data. A benchmark is a computer program that represents a company's primary workload or proposed needs. It provides a standard against which a new product can be measured. A workload sample is taken to an installation where the vendor's product is being used, and actual measures of how well the product handles the workload are obtained to make relative comparisons of performance. Users can evaluate test results to determine which hardware device or software package displayed the best performance characteristics. Benchmark tests are good and efficient, but they are expensive. And it is difficult to arrive at a representative benchmark and the programming required to run it.

Fourth, special simulators can be developed that simulate the processing of typical jobs using the products of the vendors who have submitted bids. Their performance is then compared and judged.

A final option is to use a scoring system of evaluation when there are several competing proposals for a hardware or software acquisition. Each evaluation crite-

TABLE 10-1 Scoring system for CPU evaluation.

Factors	Weight (Maximum Score)	1	2	3	4
CPU					
Addressing	30	22	18	26	26
Arithmetic capability	20	14	12	10	12
Communication ports	40	38	17	31	17
Cycle time	30	16	22	28	16
I/O channels	50	42	34	18	26
Product life cycle	40	36	36	29	26
Registers	30	22	10	24	18
Word size	50	50	30	10	10
Subtotal		240	179	176	151
VENDOR					
Business position	15	12	14	12	12
Delivery time	40	24	31	38	38
Location	20	14	10	18	18
Maintenance	25	19	16	13	22
Unit installed	30	22	26	14	26
Performance history	30	22	22	18	18
Training offered	30	30	25	15	20
Subtotal		143	144	128	154
Total		383	323	304	305

rion is given a certain number of maximum points. Each competing proposal is assigned points depending on how well it meets the specifications of the prospective user. Scoring each criterion for several proposals helps to organize and document the evaluation process by spotlighting the strengths and weaknesses of each bid. Table 10-1 shows an example of a scoring system.

Negotiating Contracts

The final phase of the acquisition sequence is contract negotiation. After requirements have been determined, RFP has been prepared, bid solicitation has been offered, and bid evaluation complete regarding a decision on the vendor selection, it is time to negotiate the contract.

Points that should be negotiated and made clear in a contract include delivery, testing, acceptance, installation, support during and after installation, and termination. Two critical points to be considered in contracts are backup provisions and client rights regarding future product modification or enhancements. The more specific the contract, the less likely for problems to arise in the future. It is advised to seek legal counsel before signing the contract (preferably a counsel with experience in computer related contracting).

Financing Acquisitions

There are three financing options for CIM-related computer hardware—leasing, purchasing, or renting. Software can be developed, leased, or purchased. There are advantages and disadvantages of each method of financing. It is important to consider the method that will best meet the needs and goals of the organization.

Leasing

A major development occurred at the time of the third generation computers: leasing CIM system hardware. An independent computer leasing company is generally involved. Leasing companies purchase equipment and then lease it to users for a long period—usually two to five years. Leasing arrangements typically include a maintenance contract, purchase and trade-in options, and no charges for extra shift operation. However, a cancellation penalty is charged if the company terminates a lease before the end of the assigned minimum lease period. Leasing does not require the financing of a large purchase price and is less expensive than renting equipment for an equal period. The major disadvantage is the long-term period of the lease contract that cannot be terminated without the payment of a substantial penalty.

Purchasing

Many companies using microcomputers, minicomputers, and related equipment for CIM are now opting to purchase the equipment because the prices are low enough to make purchases affordable. Purchase has a tax advantage because buying a computer is considered a capital investment. This allows computer users to qualify for an investment tax credit, which reduces their income tax liability. The major disadvantage of the hardware purchase is that equipment maintenance is not included in the purchase price and must be arranged separately with the computer manufacturer or an independent maintenance company.

Renting

The rental price generally includes the cost of maintenance, and the rental agreement can usually be canceled without penalty by the user with only a few months' notice. Renting provides greater flexibility in changing equipment configurations, and it greatly reduces the risk of technological obsolescence because users are not locked in. The major disadvantage of equipment rental is the higher cost incurred if equipment is rented for more than four or five years. Prices are usually based on a two- to four-year life, during which the rental agency will recover the cost of the equipment with a substantial profit. Therefore, if the computer is going to be used for a longer period (longer than 176 hours per month), the cost of rental is higher than the cost of purchase. This consideration should be used by the company as a reference point to make the decision whether to purchase or rent.

Testing

The implementation of CIM requires that the new system be tested, which not only involves testing and debugging all computer programs but all components of the system. Components that must be tested include equipment (old and new), new forms, new software, new data collection methods, new work procedures, and new reporting formats.

The components can be tested independently of the system as well as parts of the total system. Tests give information about accuracy, range of inputs, frequency of inputs, unusual operating conditions, the characteristics of human errors, and reliability of the system. Difficulties occurring during testing of components lead to design modifications that will provide benefits when operations are carried out.

An interesting point is that testing shows only the presence of errors, not their absence. It is through design that error-free products can be produced.

Testing occurs not only in the implementation phase but throughout the development process. For example, input documents and procedures are tested before the final form is determined to allow users the opportunity to suggest modifications. As each part of the system is installed, tests should be performed in accordance with the test specification established during the design phase.

Testing the system is an important part of the implementation phase because it gives the supplier the opportunity to increase the likelihood that his products will perform as specified in the designs. Testing considers all possible conditions or situations that might arise, capacity processing loads, and linkages with other systems by compiling representative data to be tested by the product. Test data are an invaluable outcome of the test process. Some test data are generated automatically by software; other test data are considered "live" (i.e. taken from the organization's current transaction files).

Training

In addition to selling management, users, and operating personnel on the CIM concept, training involves their education in the specific skills needed to operate and use the system. If training is adequately provided, the shock effect of transferring to a new system will be minimized. Training should not be a one-time event but should be an organization-wide, ongoing program that supplements the initial training as needed over the life of the project.

Because CIM is an integrated environment, it affects many people. The success or failure of the system can hinge on the strength and adequacy of the training along with the support given by top management. All persons involved must be trained concerning (1) how the system will help them, and (2) what their role is in the system.

Who Needs to Be Trained

The internal groups who need to be trained include management (on all levels), operations, clerical support, and users outside of the computer department. External groups consist of vendors, customers, and labor unions.

Until recently, training often carried a low priority. Companies seemed to think that employees should come prepared for the job, needing only small amounts of on-the-job training for specific tasks. With CIM coming into the picture, that attitude is changing. By the year 2000, the American Society for Training & Development (ASTD) predicts that 75 percent of all workers currently employed will need retraining because they do not have the skills required for the CIM environment.

Training Methods

A number of training methods can be used to effect CIM training. A combination of methods is usually the best approach.

The training method selected will be determined by the kind of training or learning you want to achieve. In a training setting for CIM, the most appropriate methods may include hands-on, lecture, or video. Any one or a combination can be considered.

Hands-On Training

Hands-on training includes instructions on how to perform a specific task and how to provide the opportunity for the trainee to perform the task in a training environment. Groups needing this type of training include production employees and the information services staff.

Lecture Method

The lecture method of training includes instructions on how to interpret the benefits of the system and the duties and responsibilities of the individuals in the system. It provides the opportunity to gain information on the operation of the system. Groups needing this kind of training include information service staff, production employees, clerical employees, sales representatives, management, customers, and organized labor. The lecture method is often used to introduce management to new concepts.

The lecture method is not perceived to be the ideal training method, but it does provide the opportunity to share a great deal of information with a large number of individuals.

Video Training

Video training can include all the instructions for information services staff, production and clerical staff, and organized labor. Video training has the disadvantage of not having feedback or interaction with the participants. It can be an effective method in combination with other methods, however.

Simulation Training

Simulation training instructs the students how to perform specific tasks under particular conditions. A situation is then modeled for the participants so that they can make decisions or take actions on the basis of the training they have received. Simulation training is beneficial for operational, tactical, and top-level management groups.

Classroom Training

Classroom training can be provided off-site at a college, university, or training center. All other methods of training can be in combination in the classroom environment. The classroom method is particularly suited to the presentation of concepts and the sharing of ideas.

Installation

The time when the new component replaces the old component or the new system replaces the old system is the conversion phase.

The probability that some failure will occur with any CBS implementation is high. Even though the system will have been tested, there are still likely to be errors in it. Sometimes the system fails because of a change in the original specifications. Such changes may be caused by a time lag between when the specifications were devised and when the system was introduced. Organization changes may also have taken place during the time lag. The more sophisticated or innovative the system, the higher is the probability of failure.

Another problem that may occur during conversion can result from inadequate training of personnel. The problem arises when people are suddenly thrown into a new situation with new procedures, new equipment, and new co-workers. If the training has not been adequate, mass confusion may result. The choice training method is critical in reducing the risk of a disastrous failure.

Installation Methods

The most common methods of installation are crash, phased, parallel, and pilot (see Figure 10.3).

Crash

In the crash method, the old system is discontinued at a certain point, and the new system is still not in operation. This method is the least expensive and time-consuming, but it carries the greatest risk of problems. This method may work in small companies but can create severe problems in large ones.

Phased

In the phased method, the new system is incrementally introduced by dividing it into subsystems and introducing each subsystem only after knowledge and experience are gained from the previous one. This process prolongs the installation period.

CRASH

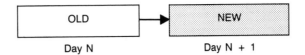

PHASED

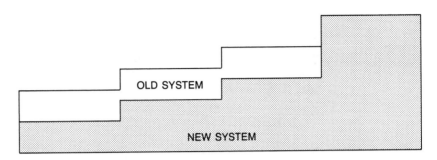

PARALLEL

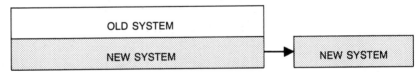

Figure 10.3 Installation Methods

An example of the method would be to install the order entry subsystem first, follow it by the production segment, and so on. Or the installation of the subsystems could proceed by geographic location.

Parallel

In the parallel method, two systems operate simultaneously for a period. Its greatest advantage is that it allows the new system to be fully debugged, using live data, before the old system is discontinued. This method has the greatest potential for reducing the risk of failure, but it is the most expensive because of the cost of operating two systems simultaneously.

Pilot

When the pilot method is used, one subsystem of the new system is introduced and experience is gained from its use, after which the remainder of the system is introduced.

The selection of the installation method depends on urgency, importance, size of organization, extent of organization dependence, required human interaction, and budget. When the system is complex, it is common to have mixed installation methods in which particular subsystems are introduced according to their unique attributes.

Security and Control

Control is the process used to regulate and guide the operations of a system. It should begin with the planning phase and continue throughout the development cycle. This responsibility of control is a management function.

Control procedures should guide and regulate backup and recovery. In addition, there should be security to guard against the modification, destruction, and disclosure of data, along with security measures against natural and human hazards. It is difficult, however, to make a system 100 percent secure.

Control Components

According to McLeod [1], an information system should have three properties: integrity, auditability, and controllability. For a system to have integrity, it must perform according to specifications. The system should be *fail-soft,* which is the ability to continue performance even if one or more components fail.

For a system to be auditable, it must be easy to verify its performance. An identifiable individual should be responsible or accountable for the performance of every component and event in the system.

For a system to have controllability, each subsystem should have independent functions. In this way, should something happen in one subsystem, the total system will not be compromised. Control should be established and maintained for four areas—source document, input, processing, and output controls. Source document controls can include sequential numbering of documents, time stamps, and batch totaling to provide checks on the system.

Input controls should include measures to assure accurate conversion of source documents to machine-readable form. Source data automation techniques and controls are means of providing accurate machine-readable input data. These techniques provide for the data to be converted at the site or in the event the data process is by direct input to the computer. Examples of this process are point of scale (POS), universal product code (UPC), magnetic ink character reader (MICR), the mark sensor, and voice recognition.

Measures should also be taken against unauthorized data input or access by using such controls as passwords, identification codes, or restricting the use of terminals to selected individuals. In addition, a good practice is to separate dates and authors to provide a check and balance.

Processing controls include those that assure complete and accurate processing of data. Measures can include an *audit trail,* which enables one to trace data or process back to the original source.

Output controls include control of the dissemination of data to authorized users. On-line output is difficult to control. If data are changed on-line, it is hard to trace; therefore, on-line programs should be logged to the transaction on computer tape. These logs can be audited for accuracy and error correction.

Maintenance

Specifications can change during the lifetime of a system. New functional requirements will be needed, and old functional requirements will need modifications or replacements. Changes in the external environment will require corresponding changes to the system model. These situations require system maintenance. Maintenance is used here to denote altering the system to meet these changing specifications.

Maintenance is an ongoing activity that keeps the system at its highest level of efficiency and effectiveness within the constraints of the organization. The primary focus is to reduce errors caused by the design or environment while improving the scope and services. Maintenance functions in CIM include routine preventive operations as well as overhaul and repair.

Maintenance Consideration

The internal and external environment affects maintenance. The environment must be monitored closely to provide the necessary maintenance.

Internal environment factors that can affect maintenance are no provision for a maintenance program, no budget for maintenance, no understanding of the importance of maintenance, no cooperation from users, no documentation, and no qualified personnel. These factors are sometimes the result of a flawed systems design, which has a rippled effect into implementation. These areas were discussed in more detail in Chapter 9.

External environmental factors that can affect the need for maintenance are government policies regulations and legislation, economic conditions, competitive conditions, and new technology.

Government Policies

Changes in government policies, regulations, and legislation require compliance by the organization. For example, changes in pension rules and financial disclosure require updating the system. Such updating is considered to be maintenance.

Economic Conditions

Changes in economic conditions, such as local or national unemployment rate, inflation, and interest rates, can have a great impact on the organization's environment. It is difficult to predict these conditions; as part of a periodic maintenance program, such conditions should be evaluated regularly.

Competitive Conditions

Changes in competition in an industry, such as the expansion or collapse of a product line, a new price policy, new standards, and alternative products, can have a financial impact on the organizational environment. Conditions such as these can require major changes in the maintenance program to modify the objectives of the organization.

New Technology

Changes in technology require continual system maintenance. A technology not available last year or even last month may now be available. The procedures needed to incorporate it into the system to make it more effective and efficient are part of maintenance.

Maintenance Methods and Techniques

Good maintenance techniques include preparing a log for change requests, prioritizing all requests, making monthly and annual plans, and documenting maintenance as an event occurs. Typically, one-third of the total system budget is allocated for maintenance activities, which in itself establishes the importance of the procedure.

Documentation

Documentation is the preparation of written descriptions of the scope, purpose, information flow components, and operation procedures of the system. It is an integral element of the system because it provides both visibility and understanding of the system, product, or service. Documentation provides a medium by which a common view of the participants in the system can be obtained.

Documentation is a necessity for troubleshooting, replacing subsystems, interfacing with other systems, training new operating personnel, or evaluating and upgrading the system. Documentation communicates the activities and results of each stage of system development. Proper documentation allows management to monitor the progress of a project to minimize problems that arise when changes are made in the design of a system. Installing and operating a newly designed system or modifying an established system requires communication about the previous operation or about the design and requirements of the new one.

What Should Be Documented?

What should be documented is always a basic question when documentation is discussed. To answer the question, one must ask: If the system is properly documented, what outcomes can be expected? Proper documentation provides the following:

1. A valuable data source for developing schedules, costs, and staff plans.
2. A common reference for managers, designers, and programmers for system maintenance.
3. Enough information to enable a person not familiar with the design to reconstruct it.
4. A means of training for new personnel.

Documentation Methods and Techniques

Methods of documentation include provision of a user's manual, operator's manual, project manual, and procedures manual.

A user's manual describes the applications of the system, product, or service. It should describe the purpose, environment, benefits, limitations, job description, and contract. A clear, correct, concise, and complete user's manual is imperative because it is through this documentation that the system can be used. If potential users are not supplied adequate support through documentation, the system may fail.

An operator's manual gives the information needed for the scheduling, preparation, and execution of a run. It also describes the input, the process, and the output of the product or service from the system along with the files, hardware, and software to be used. The manual should be continually updated as the system grows.

A project manual describes all points in a project, including a description of the product, detail of the project's history, name of the project leader, current and future status, constraints, objectives, evaluation, and major decisions and how they were reached. A project manual is critical in maintaining the project for training new personnel on the project.

A procedures manual describes procedures in operation, recovery, error conditions, and data entry, processing, and retrieval. It also provides a summary of the system and organizational requirements, an overview. A procedures manual is sometimes called a system manual. It provides examples of forms and reports, an index of computer programs, and the database definition, along with hardware and software specifications.

Evaluation

After CIM has been in operation for a short time, each step in the design of the system should be evaluated. Evaluation is the follow-up to the design and implementation process. It is considered to be part of the implementation because it evaluates the new system. Evaluation analyzes the efficiency and effectiveness of the system. The primary factor to consider is the effectiveness—that is, the extent to which the goals and objectives of the system have been met.

Exercises

1. Explain what system development is and discuss the components of the system development cycle.

2. List and briefly discuss the implementation components of a system.

3. Explain why a Gantt chart may be helpful in the planning phase of the implementation process.

4. Develop a Gantt chart for tasks that would be involved in implementing a system (e.g., CIM, telecommunication, security, inventory).

5. Obtain a copy of a RFP form from a business or industry and complete one for a new acquisition for your department's new system (ask your instructor what the needs are in the department). Make sure the information you use on the form gives the actual specifications for equipment.

6. Devise a scoring system for evaluation of the selection of the new acquisition proposed in Exercise 5.

7. List and briefly discuss the three methods of financing acquisitions. Include in the discussion the advantages and disadvantages of each method.

8. Develop an outline of the components that it would be necessary to include in a training program for a CIM system. Include the who, what, where, when, how, and why of the training program that is to be developed.

9. Discuss the common installation methods and when to use each method.

10. Explain what fail-soft means in connection with security and control.

11. Explain how source data automation contributes to control of a system.

Self-Study Test

1. The system development cycle begins with a concept to meet the needs of the organization.
   True   False

2. System development is not the umbrella of the development process.
   True   False

3. Implementation is the framework to perform the integration of product, service, or system.
   True   False

4. Planning is the third step in the implementation process.
   True   False

5. Involvement of top management is not considered necessary for the successful implementation of a system.

    True   False

6. The determination of a CIM system requirements show the importance of the acquisition process.

    True   False

7. A benchmark is a computer program.

    True   False

8. Contract negotiation is the final phase of the acquisition sequence.

    True   False

9. Three methods of financing computer hardware are leasing, purchasing, and renting.

    True   False

10. Testing a CIM system is the orientation and selling to management, users, and operating personnel the CIM system.

    True   False

11. Those involved in a system must be trained on (1) how the system will help them and (2) what their role is in the system.

    True   False

12. The type of training method used is determined by the kind of training or learning you want to achieve.

    True   False

13. When the new component replaces the old system or the new system replaces the old system, this is considered the maintenance phase.

    True   False

14. The more sophisticated or more innovative the system, the higher the probability of failure.

    True   False

15. It is not unusual to make a system 100 percent secure.

    True   False

16. The internal and external environment affects maintenance.

    True   False

17. Five methods generally used to assess vendor claims for their products include all except:

1. Trial periods
2. Benchmark testing
3. Leasing
4. Simulation modeling
5. Scoring

18. To share a great deal of information with a large number of individuals, the _____ training method is used.

1. Simulation
2. Video
3. Lecture
4. Hands-on
5. On-the-job

19. _____ training is beneficial for groups on the operational, tactical, and top-management levels.

1. Simulation
2. Video
3. Lecture
4. Hands-on
5. Classroom

20. A method of system installation is:

1. Crash
2. Phased
3. Parallel
4. Pilot
5. All these.

21. There should be _____ against data modification, destruction, disclosure, and human hazards.

1. Cost
2. Security
3. Execution
4. Integration
5. None of these.

22. In relationship to control, information systems should have:

1. Planning, training, and integrity
2. Maintenance, documentation, and cost
3. Planning, testing, and auditability
4. Integrity, auditability, and controllability
5. None of these.

23. _____ is the ability to continue performance even if one or more components fail:

1. Integrity
2. Fail-soft

    3. Fail-safe

    4. Controllability

    5. Security

24. If a system has _____, it should be easy to verify its performance.

    1. Auditability

    2. Controllability

    3. Integrity

    4. Security

    5. Accuracy

25. If a system has _____, each of its subsystems should have an independent function.

    1. Auditability

    2. Controllability

    3. Integrity

    4. Security

    5. Accuracy

26. _____ allows data to be connected at the site or event where the data take place with direct input to the computer.

    1. Audit trail

    2. Security

    3. Source data automation

    4. Conversion

    5. None of these.

27. _____ enables input data or the process executed to be traced back to the original source data.

    1. Source data automation

    2. Audit trail

    3. Training

    4. Fail-soft

    5. All of these.

28. _____ is an ongoing activity to keep the system at its highest level of efficiency and effectiveness within the constraints of the organization.

    1. Maintenance

    2. Documentation

    3. Testing

    4. Security

    5. All of these.

29. External environmental factor(s) that can affect the need for maintenance are:
1. Government regulations and new technology.
2. Economic and competitive conditions
3. Government regulations only
4. No documentation of qualified personnel
5. Items 1 and 2.

30. One of the following is an integral element of the system because it provides both visability and understanding of the system.
1. Maintenance
2. Documentation
3. Teaching
4. Planning
5. Testing

ANSWERS TO SELF-STUDY TEST

1.	True	**7.**	False	**13.**	False	**19.**	4	**25.**	2
2.	False	**8.**	True	**14.**	True	**20.**	5	**26.**	3
3.	True	**9.**	True	**15.**	False	**21.**	2	**27.**	2
4.	False	**10.**	False	**16.**	True	**22.**	4	**28.**	1
5.	False	**11.**	True	**17.**	3	**23.**	2	**29.**	5
6.	False	**12.**	True	**18.**	3	**24.**	1	**30.**	2

References

[1] McLeod, R. *Management Information Systems.* Chicago: Science Research Associates, 1986.

Index